Routledge Revivals

SAHARA
THE GREAT DESERT

SAHARA

THE GREAT DESERT

BY
E.-F. GAUTIER

AUTHORIZED TRANSLATION BY
DOROTHY FORD MAYHEW

WITH A FOREWORD BY
DOUGLAS JOHNSON

Routledge
Taylor & Francis Group

First published in 1935 by Columbia University Press

This edition first published in 2018 by Routledge
2 Park Square, Milton Park, Abingdon, Oxon, OX14 4RN
and by Routledge
52 Vanderbilt Avenue, New York, NY 10017, USA

Routledge is an imprint of the Taylor & Francis Group, an informa business

© 1935 by Taylor and Francis

Publisher's Note
The publisher has gone to great lengths to ensure the quality of this reprint but points out that some imperfections in the original copies may be apparent.

Disclaimer
The publisher has made every effort to trace copyright holders and welcomes correspondence from those they have been unable to contact.
A Library of Congress record exists under ISBN:

ISBN 13: 978-0-367-18265-6 (hbk)
ISBN 13: 978-0-367-18266-3 (pbk)
ISBN 13: 978-0-429-06038-0 (ebk)

SAHARA
THE GREAT DESERT

Photograph by Gautier

A KHOTTARA IN THE OASIS OF TIMMUDI, LOWER BASIN OF WADI SAURA

SAHARA
THE GREAT DESERT

BY

E.-F. GAUTIER
Professor in the University of Algiers

AUTHORIZED TRANSLATION BY

DOROTHY FORD MAYHEW
Librarian, Institute of Geographical Exploration, Harvard University

WITH A FOREWORD BY

DOUGLAS JOHNSON
Professor of Physiography in Columbia University

NEW YORK : MORNINGSIDE HEIGHTS
COLUMBIA UNIVERSITY PRESS
1935

SAHARA: THE GREAT DESERT is translated from the second French edition of E.-F. Gautier's LE SAHARA (Paris, Payot, 1928) and from hitherto unpublished material supplied by the author. To the photographs and line drawings reproduced from the original edition, there have been added 27 drawings and a map, all by PAUL LAUNE.

PRINTED IN THE UNITED STATES OF AMERICA
QUINN & BODEN COMPANY, INC., RAHWAY, N. J.

AUTHORIZATION

Dear Miss Dorothy Ford Mayhew:

Your translation of my *Sahara* is admirable for intelligence and faithful adaptation. It has my entire approbation. And I am happy to express my thanks publicly.

Very sincerely yours,

(Signed) E.-F. Gautier

FOREWORD

In February, 1924, the writer of this Foreword found himself on a small steamer en route from Marseilles to Algiers. The mistral, sweeping out of the north, was lashing the blue waters of the Mediterranean into fury. Under such conditions one is apt to regret having essayed a winter crossing, unless the objective in view holds great attraction. In the present instance no discomfort could dampen my enthusiasm for a double objective: to see Émile Gautier and get a glimpse of the Great Desert which is the field of his scholarly researches.

Seated in his laboratory at the University of Algiers, I asked Professor Gautier to tell me something of his methods of exploration in the greatest of the world's arid regions. "Oh, it is very simple," he replied. "You just get some camels and some food, and go." For many years Gautier has been going; and as a result both his geographical colleagues and a far more numerous lay public have come to know the Sahara better.

I cannot pretend to introduce Émile Gautier to the English-speaking world. The eminent authority on the geography and history of the Sahara and its bordering steppes requires no introduction. Not only his books with their inimitable style, but his highly individual and wholly delightful personality, are known on both sides of the Atlantic. The walls of American classroom and lecture hall have echoed his voice; the sands of what we are pleased to call the Great American Desert have received his footprints. He knows us, and we know him; and we are friends.

I can, however, introduce to the English-reading public a new volume on the Sahara from the pen of the

distinguished French savant. The opportunity to do
this, made possible by the enterprise of Miss Mayhew
and the Columbia University Press, is at once a pleasure and an honor. When Gautier's *Le Sahara* appeared
in French, it was immediately hailed as a geographical
classic. Thanks to the labors of translator and publisher, and to the generous coöperation of the author,
English readers may now possess in their own language an expanded and revised version of this truly
admirable work.

In the pages which follow the reader will journey
with Gautier into one of the most fascinating regions
of the world. He will learn to know, as never before,
the real Sahara: the Sahara noted for exceptional
aridity, yet where exceptional rains destroy whole
towns; the Sahara where temperatures of 158 degrees
Fahrenheit may be encountered in the dune sands, and
where the traveler may find his drinking water frozen
over night; the Sahara where the Moslem Tuaregs have
a horror of ablutions, and where their nearly naked
bodies are scoured clean by sand-laden winds; the
Sahara where scarcity of vegetation gives one a
"grandiose and overwhelming impression of absolute
emptiness," and where grazing from one tiny tuft to
its nearest neighbor is "an extremely ambulatory exercise"; the Sahara where the native hunter stalks the
curious adax "for days through barren solitudes of
mortal danger," knowing that in the animal's abdominal viscera is a natural sack storing large reserves of
greenish water drinkable in extremity; the Sahara
where microbes are scarce, and serious wounds, relatively free from infection, heal with astonishing ease;
the Sahara which with its barren wastes has throughout
the centuries served as a gigantic bulkhead separating

the white Maghrib to the north from the black Sudan to the south; the Sahara across which not even the might of Rome could extend its conquering power to the southern lands beyond.

Under the expert guidance of Gautier the reader interested in the origin and evolution of the earth's surface features will discover ancient folded mountains, worn low by long-continued erosion, which constitute part of the Saharan region; and the vast plains and plateaus of horizontal rocks which make up the remainder. He will observe interesting fractures dividing this part of the earth's crust into blocks, and see some of the blocks raised and ravaged by streams to give rugged mountainous forms. He will ascend lofty volcanoes 11,000 feet above sea level, and descend into desert basins having floors lower than the surface of the ocean. He will study the work of running water in the desert, and learn how much the eroding and transporting powers of the winds have been exaggerated. He will traverse an area of sand dunes as large as France, but find that the sand has been brought to its resting place by streams, and not by winds. He will see on sandstone cliffs ancient engravings apparently as fresh as the day they were made, although subjected to "every violence of the wind for several thousands of years."

Both layman and scientist will find profit as well as pleasure in following Gautier's reconstruction of the Sahara of past geologic ages, and his discussion of progressive desiccation of the Great Desert. Some may be surprised to learn that the camel was introduced into the Sahara in comparatively recent times, just as it was still later introduced into the deserts of the southwestern United States. The consequences of such introduction were very different in the two cases, and the

American reader will therefore take a special interest in Gautier's analysis of Saharan history subsequent to the coming of this strange beast of burden. The regional geography of the Sahara has new fascination when pictured by Gautier's pen, and all who delight in eloquent descriptions of unusual places and peoples will follow the distinguished French scholar to the end of his concluding observations concerning the region to which he has devoted long years of immensely fruitful study.

DOUGLAS JOHNSON

COLUMBIA UNIVERSITY
IN THE CITY OF NEW YORK
AUGUST 1, 1935

PREFACE

THE present English version of Gautier's *Sahara* is an expanded and revised edition, containing, thanks to the author, much material which has never before been published. The conclusion, which is entirely new, gives a vivid picture of the present-day Sahara under European administration, no longer a waste region lost to the uses of the civilized world, but an open country already confronted with economic and industrial problems common to modern progress, yet showing an amazing advance and development in an incredibly short space of time. The picture is one of startling contrast with that of the prehistoric and historic Sahara so painstakingly drawn in the main body of the text, and adds vital interest to the book as a whole.

Besides this new conclusion, the author has given us copious notes regarding his own latest researches and the results of his and other expeditions which have been in the field during the past few years. He has been tireless in answering questions and furnishing explanations which have clarified many points that might have been obscure to the lay reader, all of which data has been worked into the text at various points. The map has been redrawn to include new map data, and the illustrations have been enriched by several new photographs furnished by him.

Any changes which have been made in the order of presenting material have had his consent; and the present version has been submitted to him and has received his careful consideration and approval. On behalf of the publishers and myself, I wish to express my appreciation of his generous coöperation and unflagging efforts to assist us in presenting his valuable book to the

English-reading public to the best possible advantage.

I should like also to thank the many others who by their interest and assistance have helped in this undertaking; particularly Professor Douglas Johnson of Columbia University, Professor Derwent Whittlesey of Harvard University, Professor Charles Brooks of Blue Hill Observatory, and Mr. Charles A. Weatherby of the Gray Herbarium, both of Harvard University.

DOROTHY FORD MAYHEW

CAMBRIDGE, MASSACHUSETTS
    AUGUST 15, 1935

CONTENTS

CONTENTS

PART V

CONCLUSION

ILLUSTRATIONS

ILLUSTRATIONS

PART I

INTRODUCTION TO THE SAHARA

I
The Great Desert

HE Sahara, or, as it is known in the atlases, the Great Desert, is probably indeed the most outstanding desert on the surface of the earth—not only because of its exceptional aridity, but by reason of its tremendous size as well, for it takes in the whole northern portion of Africa, nearly half a continent. Moreover, with the possible exception of the American Desert in the United States, the Sahara seems also to be one of the best known deserts of the world; or at any rate, one of the least unknown.

This knowledge has been only recently acquired, for at the beginning of the nineteenth century the Sahara was as completely unknown to the outside world as was the rest of Africa. The questions of Timbuktu, Lake Chad and the sources of the Nile were among the most important on the exploratory program, and for decade after decade still remained to be solved. The Sahara was the scene of some of the most brilliant achievements of nineteenth-century exploration. Caillé, the first European to have seen Timbuktu—Speke and Grant, who discovered the sources of the Nile—these are only some of the great names; for there was no lack of early travelers, belonging to what might be called the heroic period, whose books are still a source of information.

Particularly is this true of Barth, Rohlfs, Nachtigall, Duveyrier and de Foucauld.

This period of individual accomplishment gradually gave place to one of more comprehensive and co-ordinated effort. Saharan exploration entered into this new phase about 1880, when France in the west began by successive stages to establish her military domina-tion over that portion of the Sahara south of Algeria. The new era opened in this region with the explorations of Flatters and Foureau-Lamy. Then, beginning with 1900, a considerable number of scientific men, including officers, explorers, geologists and geographers, gathered around Laperrine, their routes crossing and recrossing, joining together and supplementing one another. It was through the combined efforts of these men that much of the almost total obscurity which so long had shrouded the Occidental Sahara was cleared up. At the same time the Geographical Service of the Army, although ab-sorbed in Algeria, Tunisia and Morocco, was also de-voting an increasing activity to the desert.

During the same period, England to the east had established herself in Egypt; and the British Geolog-ical Survey of Egypt was undertaking an investigation of the Oriental Sahara, which has resulted in the pub-lication of many valuable maps and monographs on that region, particularly on the oases of the Libyan Desert.

The part which has been least studied is the Central Sahara. Italy, now in control of it, did not enter on the scene till very late, just at the eve of the Great War. She has consequently not had the time, and still less, on account of the War, the leisure, necessary to organize a scientific conquest. She has, however, made a begin-ning. Furthermore, this is the region on which we have the books of Barth, Nachtigall and Duveyrier; while the two Tilho expeditions, going in from the French Sudan, have definitely solved the Chad question, and the second of these has given us data on the Tibesti area. The mysterious Kufara has also been thoroughly documented since information completing that of Rohlfs has been brought back to us by several more recent travelers, notably Lapierre, Rosita Forbes and Hassanein Bey. Moreover, according to our latest information (1933-34), there are now Italian posts distributed throughout the length of Tibesti, adjoining the French posts to the south; while the oasis of Kufara itself has been occupied by an Italian garrison and is already linked to the Mediterranean by an excellent motor highway.*

In view of all this, it seems permissible to say that the Sahara is at present better known than most of the other extensive deserts of the world; in any case, it is certainly possible at this stage of our knowledge to attempt a picture of the whole.

EXPLANATION OF THE DESERT

In the presence of these vast sterile expanses, sprinkled with salt and strewn with dunes, certain portions of them lying below sea level, the first impression, and one which long persisted in men's imaginations, was that we were dealing with a dried-up sea bottom. This was nothing but a popular notion. For a desert is a continental surface like any other; the explanation of its aridity is not furnished by its geological past. It is arid because it has not sufficient rainfall; because the amount of water which it loses by evaporation is not equalized by that which falls from the sky.

* See *Le Sahara Italien*, official guide to the Italian section of the Saharan Exposition at the Musée du Trocadéro [Rome, Colonial Ministry, May, 1934].

It is also generally agreed that the distribution of deserts over the earth's surface is a phenomenon which is exclusively climatic; and we likewise know that climate is first of all a function of latitude. Now we find that the arid zones on the continental surfaces in both hemispheres are interposed between the equatorial and the temperate zones; while extending over the oceanic surfaces in these same latitudes there are belts of atmospheric high pressure which separate regions of relatively lower pressure on either side: those of the easterly trade winds in the tropics from those of the westerly winds in the higher latitudes. Thus in almost every case the desert regions on land correspond roughly with these bands of oceanic high pressure. This is a general rule, applying to the deserts of North and South America, to the Kalahari and the Australian desert, and even on a smaller scale to the subdesert region of Madagascar.

To consider this in more detail, let us glance at a chart of the Atlantic Ocean showing the isobars, or lines of atmospheric pressure. In an almost exact prolongation of the Sahara will be found the "Azores maximum," a belt of barometric high pressure (30.3 inches) which stretches across the ocean. Above this belt the North Atlantic is swept the entire year by cyclonic depressions, a few of which, with their attendant precipitation, are occasionally able to reach Africa, either direct from America or by way of Europe. The latter are of the greater practical importance and are known as Gulf of Genoa depressions, being "lows" of North Atlantic origin which have been broken over western Europe and regenerated over the Gulf of Genoa. It is largely owing to their agency that eastern Algeria and the northern boundary of Tunisia are somewhat more rainy than western Algeria and Morocco. These occurrences take place only during the winter, however, in the season when the Azores "high" is farthest to the south.

On the other side of this high-pressure belt lies the region of the tropical rains which annually follow the sun on its journey north and south, taking the form of violent downpours that break in the afternoon. On the continent of Africa these rain storms occur in Senegal, where they are known as "tornadoes"; but they seldom go any farther to the north, and consequently have little effect on the desert climate.

Thus with the North Atlantic depressions to the north of it and the tropical rains to the south, we find the Sahara stretching across the continent more or less on a line with the Azores high which extends over the ocean. In the present state of our meteorological knowledge, we cannot make any definite statement regarding the atmospheric pressure in the Sahara itself. But that there is some relationship between the desert and the oceanic high-pressure belt is evident; for it is too constant to be a fortuitous coincidence.

Latitude however is not the only factor which determines climate. There are two others of great importance which may either reinforce its influence or, by acting in opposition to it, greatly modify the climate of a given place. These are the formation and the altitude of the emerged lands. Let us consider how these two factors operate. We find that in South Africa, as well as in the two Americas, the continental coast runs north-south, at right angles to the latitude; and in the Americas the coast is skirted closely by powerful mountain ranges lying in the same direction; while in Asia the influence of the latitude is opposed by the highest and most massive ranges on the globe.

In the Sahara, the contrary is true in every respect. Along the Mediterranean the North African coast lies in the same direction as the latitude, prolonging the axis of the oceanic barometric high for nearly 2,500 miles in an almost straight line. Likewise, while the Sahara does have its mountains, there is nothing which can be compared in height to Tibet and the Himalayas, nor to the Rocky Mountains or the Andes. Neither is there any continuous mountain barrier. In its general make-up, low plains or those of very moderate altitude predominate. Both these facts are of great significance, and the latter is particularly important from the barometric and thermometric points of view.

Thus we see that whereas the climatic influence of the latitude is elsewhere resisted either by continental direction or by high elevations, or by a combination of the two, in the Sahara both the direction of the coast and the moderate height of the land tend to exaggerate this influence. And there, doubtless, to put it in a very general way, we have the reason why the Sahara breaks all records as a planetary desert.

II

Climate

THE essential characteristic of the Sahara being its climate, obviously we should give a long and detailed description of it. Unfortunately for our purpose no thorough scientific investigation has as yet been made. The Meteorological Service of Algiers has rather recently installed an excellently equipped branch station at Tamanrasset, in the Ahaggar region, but this has been in existence too short a time to be of much help to us; and, in spite of the relatively large number of haphazard observation posts scattered throughout the desert, there is nowhere else a really up-to-date and well-organized station for systematic research. Lacking, therefore, a base of reliable information, we must resign ourselves to making a more or less literary study of the climate, insufficiently supported by precise observational data, and without the usual diagrams and statistical tables.

One thing which facilitates our study of the Saharan climate is its homogeneity. The desert, its greatest length extending east and west, lies roughly between 16° and 29° north latitude, which means that it is traversed through the middle, from end to end, by the tropic line. Aside from the few isolated points which attain an altitude of some 10,000 feet, its climate is practically uniform throughout. There is, for instance, no essential difference of an atmospheric nature between the Egyptian Desert and the desert region in French Territory.

The most significant factor in the desert climate is the rainfall, since it is the insufficiency of rain that brings the desert into being. We must make clear at the outset that we are not dealing with an absolutely rainless region, for there is no one spot on the whole globe where there is not some rainfall, be it much or little; and the Sahara does have its rains. What we are concerned with, then, is to determine the amounts and frequencies, and in this respect particularly we are hampered by the lack of adequate data from the meteorological stations.

From those in the French Sahara we have figures which, if considered hastily, would seem to show an average rainfall varying around 3.93 inches, rather less than more. In order to get a true picture of the situation, however, we must study the individual records, such as those from Tamanrasset. Here we find in the year 1910 an absolute zero for every month of the year; while in 1922 the records for January show the following item:

On January 15th, at 8 P.M., a hurricane broke over the region, followed by a torrential rain. The roofs of the houses almost all fell in, and the native population took refuge in the *burj* and in the fortress as the waters carried away the small dwellings and gardens bordering the wadi. Rain continued to fall on the 16th and the wadi overflowed, the water racing past with the speed of a horse gallop. At 5 P.M. the outer wall of the fortress collapsed, burying twenty-two persons. With the icy rain still falling, the victims were dug out; there were eight dead and eight injured. The rain fell less heavily on the 17th; the wadi subsided and the weather cleared. There was seen to be snow on the neighboring summits.

From the eastern Sahara we get similar records. At Cairo only eighteen rainfalls of more than 0.39 inches were registered between 1890 and 1919, and in seventeen years out of the thirty they were entirely lacking. This includes the whole series of seven consecutive years from 1909 to 1916. But in contrast to this, the rain gauge at Ezbekieh, on January 17, 1919, registered 1.7 inches at a single fall. There were boats in the streets of Cairo, the tramways were sunk in mud to their windows, and in the Manchiet el Sadr quarter the houses of unbaked brick melted away like lumps of sugar.

It is this irregularity that particularly characterizes the Saharan rainfall, and it accords with the experience of all who have lived in the desert. The author, during a stay of eighteen consecutive months in the French Sahara, never witnessed a single heavy rain. Yet there is not an oasis in any part of the desert where the inhabitants do not retain a clear remembrance of their own last great storm and the havoc caused by it. These rains always wreck the towns, since most of the little houses and the inclosing walls are not only constructed of dried mud, but often of a salt mud; and under the deluge they melt and crumble. But this is not important, for the damage is easy to repair. They gladly resign themselves to the task because these rare heavy, devastating rains are the only ones which have any practical value; they alone feed the subterranean reserves of water and have an agricultural importance. The little showers, on account of the rapid evaporation, return almost instantly to the heavens whence they came. The M'zabites at one time threatened to emigrate from their oases because in twelve years they had not had a really heavy rain.

The Sahara, in short, is a region which has no season of regular, annual and general rains. All of its abundant and useful rains it owes to the passage of storms, the dates of which are entirely erratic, and the effects more or less confined to one locality.

Perhaps the fundamental characteristic of the desert climate may be considered the dryness of the air. To quote a few figures that will give the imagination something to grasp, let us again refer to those of Tamanrasset for 1910. This station is in the heart of the French Sahara, in the Ahaggar massif at an altitude of 4,600 feet, and far from any extensive oasis where the irrigation flow might influence the hygrometer. The relative humidity there varies from month to month between 4 and 21 percent, and the absolute humidity, or weight of water vapor per cubic foot of air, between 0.4 and 1.6 grains.* The evaporation in such an atmosphere must obviously be very great, especially as the Sahara is one of the regions of the world which is subject to the most intense heat, the thermometer reaching maxima of 122° Fahrenheit with a general average only a little less than this.

There are of course wide ranges of temperature. On the high Algerian plateaus, at altitudes around 4,000 feet, it is possible to cite cases where isolated individuals or even small strayed groups are known to have perished in a snowstorm. Such stories, however, al-

* The monthly average of relative humidity in New York City ranges from 65 to 72 percent; the absolute humidity from 1.4 to 6.5 grains per cubic foot. For London the range is between 73 and 86 percent in relative humidity, with an absolute humidity of from 2.4 to 4.66 grains.

Air Photograph

WAVE FORMATIONS OF THE SMALL DUNES

though authentic enough, are not to be considered here. For the African desert is not the Siberian desert; its blizzards are dangerous only in their effect of surprise, because they are exceedingly rare. Furthermore, these high Algerian plateaus are not exactly the true Sahara.

Nevertheless, snow and ice are not absolutely unknown even in the heart of the desert. Snow, although to be sure it melts usually within twenty-four hours, is occasionally to be sighted on the highest summits of the Ahaggar. Likewise, on winter mornings anywhere in the northern Sahara, when by chance one encounters a puddle of water in the vicinity of an oasis, it is frequently found to be covered with a skin of ice, cracking beneath the horse's hoofs. In fact, throughout the desert there are consistent winter low temperatures, especially at night. Tamanrasset in 1910 had fourteen days of frost, with absolute minima of 19.6° and 28.4° F. in January and February; while the stations at Adrar and In Salah, only four or five degrees north of the tropic line and at an altitude of less than 1,000 feet, have, respectively, seventeen and nine days of freezing temperatures a year, the absolute minima not going below 26.5° F. These almost rigorous cold spells have considerable practical importance, in that they make possible the cultivation of certain kinds of dates, which happen to be the most marketable. They are also in some measure responsible for the fact that, in the diffusion of the human species, the Sahara has become the home of a white race.

Even more remarkable than these seasonal variations of temperature are the abrupt and extreme diurnal fluctuations which are peculiarly characteristic of the desert, especially of the erg, or sandy desert, where we find the great masses of the dunes. These immense accumulations of sand do not react to the heat of the sun in the same fashion as the rest of the ground. At the surface of the dunes the mobile grains of sand touch only in limited portions of their periphery; air is imprisoned between them, forming an insulation against the heat. The heating thus remains localized to the surface, which becomes scorching hot.

It is in the sand of the dunes that such abnormal temperatures as 158° F. have been observed. In the battle of Metarfa, which was fought in the dunes, the native foot soldiers, incapable of holding a prone position for firing, remained standing in spite of orders and were all killed. Even an animal cannot endure these terrific surface heats, for it is said that in the dunes of Gurara in summer a well-shod man may successfully hunt on foot; and if he raises a gazelle, has only to keep forcing it from shelter to shelter in order to bring it quickly to bay.

Alternating with these daytime and summer heats are the nightly and winter low temperatures. At the Tinoraj well in the Erg-er-Rawi, on the twenty-fifth of February at six o'clock in the morning, the water in a basin sunk halfway in the ground was frozen in a solid block; a tin cup caught in this ice was so firmly fixed that the whole thing, basin and all, could be lifted out of the ground by means of the cup. Early as it was, however, the rising thermometer already showed 50° F.

This nightly lowering of the temperature is very sharp, almost instantaneous, as soon as the sun sets. Even summer nights in the dune are delightfully cool. If, on the other hand, after a stifling day one camps at the foot of a rocky wall, the overheated stone continues for a long time to give out warmth, making the first hours of the night very painful. We might say that the

two types of Saharan surfaces, sandy and rocky, react
to a certain extent in much the same way as do the continental and oceanic surfaces of the earth as a whole,
the former in each case being much poorer conductors
of heat than the latter.

Now if we consider the enormous dimensions at-
tained by the great dune masses, it seems quite probable
that the sharp fluctuations of temperature in these large
areas should in some way affect the distribution of
barometric pressures. This in turn must undoubtedly
have some influence on the direction of the winds. We
might expect the wind direction in the desert to show
the effect of some cause that is not felt elsewhere; and
certainly the Egyptian khamsin, for instance, to judge
from its direction, would seem to have its origin in the
great erg of the Libyan Desert.

The wind itself partakes of the violence of the climate. We must not put too much faith in the popular
exaggerations, particularly disordered as they are in
oriental countries; stories of caravans annihilated by
the simoom belong strictly in the realm of the legend-
ary. But the violence of the wind is certainly one of the
most characteristic traits of the desert, a violence doubtless enhanced by the fact that there is no mantle of vegetation to break its force. The wind is the very life of
the desert.

It blows in gusts and swirls, is particularly charged
with sand and dust, and seems connected in some way
with "certain magnetic or electrical manifestations, not
yet clearly defined." It has something like an odor of
its own—at least it causes in the mucous membranes a
peculiar sensation by which its slightest breath is recognized. Essentially characterized by a dry, fiery heat,
it has a depressing effect on man and beast; and can in

certain extremely rare cases, cited as curiosities, exercise on the human organism a toxic effect. This is because the sweat glands of the body have not sufficient activity to cope with a hot, dry wind of too great a velocity, so that little by little the temperature of the skin and then of the body rises above the level of the temperature of the air. It is thus that the fatal heat stroke is caused.

The Tuaregs, good Moslems though they claim to be, have a horror of ablutions; it is like a taboo, an unavowed survival of animism. The scarcity of water is partly responsible; perhaps also the fear, verified by experience, of overexciting or slackening the action of the sweat glands. But it can be readily perceived that for a human body, exposed almost naked to the desert wind for an entire lifetime, the rites of cleanliness are superfluous; the eternal wind, charged with sand, scours the human skin and keeps it as clean as it does the slabs of naked rock on the tops of the plateaus.

The imagination of the native has played with the subject of the desert winds. To him, the little whirlwinds of dust dancing above the surface, which are not entirely unknown in any country but which are seen every day in the desert, are "waltzing jinns." The desert wind par excellence, the burning wind, in all parts of the Sahara has a name, although this is not everywhere the same. In Algeria it is the *sirocco;* in the Sahara itself, *shahali,* which signifies "wind from the south," although its direction may vary considerably from the true south; in Egypt it is the *khamsin,* the "wind of fifty consecutive days" which blows from the southwest; and it is the same one which elsewhere is called *harmattan* and *simoom.*

When the details of the climate are better known

we shall doubtless be able to trace eolian personalities more or less differentiated to accord with these various names. The khamsin, for example, which is supposed to blow without ceasing for fifty days, seems to have no exact equivalent in the Occidental Sahara, at least from the viewpoint of constancy. Nevertheless khamsin, sirocco, simoom and the rest are all close relatives, local variations of the same wind. Since this is a region of high pressures, where the air accumulates in the upper atmosphere, it is quite possible that it is a descending wind, and in this respect a distant cousin of the Alpine foehn. In any case, it is much too remarkable a personage in its own right not to arrest the attention of man in whatever part of the desert he may encounter it.

III
Organization of Desert Life

IN the absence of really precise meteorological data on which to base our picture of the Sahara, a study of the flora and fauna may give us some valuable indications as to the climate of the desert, and also its extent and limits—since the word desert, in its manifestations on this planet, is one for which we have no rigidly fixed definition. The deserts of North America and that of the Kalahari might more accurately be called steppes. Even in the Sahara we must distinguish between the desert proper and the steppes which surround it.

To the north there are certain places, as in southern Tunisia, the Syrtes and Marmarica, where the Sahara reaches right to the Mediterranean coast. But elsewhere, particularly in the Atlas region and in Cyrenaica, where the altitude modifies the desert influences, we find steppes appearing. Of these the most highly developed type is represented by the high Algerian plateaus. Likewise to the south of the Sahara we have the Sudan, a subequatorial steppe corresponding to this northerly steppe. In many other parts of the world there are similar steppes or semiarid regions, and the term desert is applied to many of them which are certainly no more barren than these. But here, where we have the Sahara proper in sharp contrast with the Algerian and the Sudanese plateaus, the difference between the steppe and the true desert is extremely noticeable.

PLANT LIFE

The flora of the northern steppe differs from that of the southern; but neither in Algeria nor in the Sudan is there ever a complete lack of vegetation. There are no immense lifeless stretches in any part of either region. But when we come to a consideration of Saharan vegetation—while we must of course stress the point that there are plants—the most impressive feature in the very great majority of cases is their entire absence. It is impossible to put into words the grandiose and overwhelming impression of absolute emptiness through which one must pass for days and days on the march.

Nevertheless, the Sahara does have its own particular flora; and while there is some tendency toward differentiation from north to south, between the Mediterranean desert and the Sudanese desert, the typical Saharan plants all have one characteristic in common. This is their ingeniousness in defending themselves against the lack of moisture. Hugging the surface of the ground away from the wind, without leaves or with very minute thorny ones, collecting their chlorophyll in fleshy branches each of which is a tiny reservoir of liquid, they are provided with roots of an unbelievable development which seek out the moisture layers to the most profound depths.

A novice in Saharan travel may be quite surprised to see his guide stop abruptly to make coffee at a spot seemingly as entirely desolate as the surrounding solitude, where his eye cannot detect a single combustible particle on the surface of the ground. In fact there is nothing on the surface to be seen, unless perhaps a tiny piece of dead stem, no longer nor thicker than the little finger. But the initiate knows that underneath there is an enormous packet of roots which will be ample to make a good-sized fire.

Each of these heroic plants can make its fight for life only if it has considerable space to send out its roots. It grows in lonely isolation. Even in the most luxuriant places there is nothing that resembles a carpet of vegetation; every tuft is perhaps fifty or sixty yards from its nearest neighbor. Grazing here from one to the next is an extremely ambulatory exercise. Nevertheless, it is such places as these that constitute the pasturages, something infinitely precious in the maintenance of life in the desert. And these pasture grounds are very scarce, being found only in rare and widely separated basins where, by exceptionally favorable circumstances, a subterranean water layer is maintained at a reasonable distance under the surface. Between two pasturages the caravan may travel not only for hours, but actually for days.

There are other localities, however, where surfaces entirely unprovided with superficial supplies of water, and ordinarily sterile, are nevertheless, after a storm, susceptible of becoming a particular kind of pasturage, known to the Arabs by a special name. These are the "ashab" pastures. The term *ashab* * does not refer to

* The ashab is defined as an ephemeral vegetation of therophytes, developing after the rains. The most abundant species is the *Savignya longistyla*, an example of what botanists call a "rosette plant." The leaves are all borne in a cluster at the level of the ground and the stems bear only flowers. The *Savignya* has several stems or long branches from 4 to 18 inches tall, and somewhat resembles the shepherd's purse except that the flowers are purple-lilac with darker veins in the petals. The seeds of this particular species are winged so that they are easily blown about by the wind, which doubtless helps to account for its prevalence in the ashab pastures. The Arabic name for this plant is "gulglan."

Three other species are mentioned by Foureau as the chief associates of

Photograph by Gautier

THE ADAX ANTELOPE OF THE ALGERIAN SAHARA

any one particular plant, but rather to a type of vegetation, the most abundant species belonging to the mustard family, but all of them having developed special tactics of their own for combating the prevalent dryness. Their survival is through the agency of their seeds, which have the faculty of resisting even extreme drought for an almost indefinite period. Should a heavy rain fall, the seed of the ashab utilizes it with an admirable speed and energy. In an astonishingly short number of days it germinates, pushes up its stem, spreads out its flowers, and forms new seeds. It knows there is no time to be lost, and it is so organized as to make the very best possible use of the exceptional godsend. Then, after a brief existence, the ashab dies; but the new seed, carried by the wind, covered with sand, wedged under a stone or in some crevice of rock, will wait, ten years if need be, for the next storm. During their short span of life however, these plants, whose every effort is expended for the purposes of reproduction, have been veritable bouquets of flowers; and these clumps of flowers are the pasturage. The camels are very partial to them, and it is a ludicrous sight to see the delicate blossoms swallowed up by their filthy jaws.

the gulglan: one is another member of the mustard family, somewhat similar in general habit and also purple-flowered, but with differently shaped leaves and without winged seeds; the second is a plantain not very different in appearance from our common weed in lawns; the third is a kind of Wandering Jew, one of the variety of desert plants which grow only under the protection of clumps of other plants. There is no evidence of structural peculiarity in the seeds of any of these species, except the ,wings of the *Savignya;* they simply have the faculty of remaining for long periods in a state of suspended animation which resists even extreme drought.

To visualize the ashab pastures, then, we may imagine a field of bare ground, sparsely occupied by a kind of purple-flowered shepherd's purse, plantains, and a small Wandering Jew. [Note by Chas. A. Weatherby, of the Gray Herbarium of Harvard University.]

Animal Life

In considering the animal life of the Sahara, we must again be careful to avoid confusion between the desert proper and the steppes. Wherever steppes are found on the surface of the globe, they have once been or still are great game preserves, stocked with vast herds of antelopes and other herbivores, huge pachyderms and wild beasts of all kinds. In this connection we may recall the prodigious bands of bison which not so very long ago roamed the American Far West. In the Kalahari too, we find indications that it also was once overrun with great hordes of ruminants; for there are certain present forms of relief in the shape of circular pools or hollows stamped out of the limestone which, according to Passarge, can be explained only as having been made by the hoofs of these animals, who, before their extermination by the European rifles, were wont to come to these spots to drink. Both these facts bear out our earlier contention that these two deserts should more acurately be known as steppes.

Today, the Sudan and the steppes of East Africa, with their elephants, rhinoceroses, hippopotami, giraffes and bands of lions, are still the big game country. The same used to be true of the high Algerian plateaus; they were the scene of elephant hunts in the days of Carthage, and furnished the circus beasts for Rome. Even as late as 1830 the French found not only the lion and the ostrich there, but the whole region swarming with native animals, of which we have a vivid picture in Jean-Auguste Margueritte's *Chasses de l'Algérie*.

But all this pertains to the steppes, and has nothing at all to do with the desert. There is a current expression, "the lion of the desert," often and justly used as a joke. There is no lion of the desert, because the poor creature would die there, of hunger and thirst. Most of the animals encountered within its confines are simply passers-by, crossing through it without lingering, by grace of admirable legs or powerful wings. This was true of the ostrich, which was occasionally found in the Algerian Sahara until the high plateaus of French Algeria became uninhabitable to him, and has since entirely disappeared from the desert as well. This also explains the grasshopper which advances on Algeria from the south, coming apparently from the Sahara; but really coming from beyond, from the Sudanese steppe.

Yet even the Sahara proper has its animal life. Whether it is a matter of fauna or of flora, we find life maintaining its struggle for existence with an admirable tenacity and ingenuity. One of the true desert animals is the adax, a species of great antelope still to be found in some places, though now unfortunately represented only by a small number of individuals, having completely disappeared from the Erg-er-Rawi, to the west of the Saura, since the day when a group of meharists, with the destructive rage of civilized man, exterminated a band of some twenty of them in a single hunt. The organism of these desert antelopes is curiously adapted to the circumstances of their environment. The adax, like the camel, has in its abdominal viscera a natural sac which serves to accumulate large reserves of water. The native hunter, following and stalking them perhaps for days through barren solitudes of mortal danger, is well aware of this anatomical peculiarity. He knows that if he can bring down one of the beasts, he will find in its entrails a provision of greenish water which, in an extremity, is at least drinkable. [See illustration, p. 20.]

In the best pasturages there are also to be found certain humbler animals, like the small gazelle and the hare. These would seem to have the faculty of going for weeks at a time without drinking, on condition of being able to browse on succulent plants. There are jackals in the vicinity of the Wallen well, where they have worn a runway in the light earth right down to the water's edge. There are also reptiles that manage to exist by spending long periods in torpor, hidden more or less deeply in the ground, among these being the big, vividly colored lizards.

In some places scarabs abound on the caravan trails, attracted by the camel dung; and these are suspected of being able to manufacture their own water supply from the atmospheric vapors. The scourge of the Sahara is the fly, a fly both swarming and languishing; it is sucked in with the breath and swallowed, or crushed on the face, when one attempts to brush it away. But the fly belongs really to the oasis, and in the desert it travels by man or camel-back.

The flea does not exist. Microbe life is very scarce. Malaria is concentrated in the oases and is entirely unknown as soon as they are left behind. Wounds of the human body, even very severe ones, heal with astonishing ease in the Sahara, without antiseptic treatment. Rohlfs, left for dead in the region of the Saura, recovered, by the grace of God, without medical attention; and many other, though less illustrious, cases may be cited.

Human Life

Speaking in general terms, the Sahara is a region relatively azoic in character: that is, lacking in any form of life. In this respect, it is in the strictest sense of the word a true desert, even from the viewpoint of human habitation. The steppes support a specific type of humanity in the great nomad tribes who graze horses, cattle and sheep; but these powerful tribes, migratory and warlike, founders of empires, who have played so important a rôle in the history of the Maghrib and the Sudan, are not really desert dwellers.

There are some remnants of humanity clinging tenaciously to certain small corners of the Sahara, as we shall see; but their hold is a feeble one. The truly Saharan nomads are all exclusively camel herders; and while their personal energy is extreme, one characteristic common to all tribes is numerical insignificance. This has always been true, even in those rare cases when favorable circumstances and their own determination have permitted even them to play a rôle in history.

But taken all together, the historical rôle of the Sahara has been the one determined by its azoism, its sterility. Our knowledge along these lines has been considerably advanced by recent archeological activities, particularly the just-completed excavations at the tomb of Tin Hinan at Abalessa in the Ahaggar; but all the facts that we can glean tend toward the same conclusion: that the Sahara, confining its few inhabitants within their own sphere, acted chiefly as a barrier to outside influences. It arrested the explorations of the ancients as effectively as did the Atlantic. Egypt never knew the sources of the Nile. The Punic merchants of Carthage and Tripoli, as we now know, did maintain some commerce with Black Africa to the south, but it was an indirect commerce, carried on through scattered, independent desert tribes. And the great Roman Empire, although it occupied the Saharan province of Fezzan, or Phazania, never established direct relations with the Sudan.

In other portions of the earth, where the great deserts run north-south and on one side only of the continent—as in North America or South Africa for instance—the various races have intermingled and live together. But the Sahara in its whole extent forms practically a bulkhead between the blacks and the whites. The Maghrib to the north is white; the Sudan to the south is black, incontrovertibly and almost without any transition. Nor have there ever been any relations between the two, other than infiltrations, drop by drop.

It is with this azoic belt which is the Sahara proper that we are to deal in this small book. We shall not include the Maghrib and the Sudan, which are different types of regions, however interesting in their own way. There is no reason why one should not seek enlightenment on them also; but they are separate worlds, with which we are not here concerned.

Bibliography

Berthelot, André, *L'Afrique Saharienne et Soudanaise: ce qu'en ont connu les Anciens,* Paris, Payot, 1927.

Chevalier, *Plantes utiles du Sahara.*

Foureau, F., *Documents scientifiques de la Mission Saharienne,* Paris, 1905.

Gautier, E.-F., "The Monument of Tin Hinan in the Ahaggar," *Geog. Review,* Vol. XXIV (1934), p. 439.

Gautier, E.-F., and Lasserre, in *Les Territoires du sud de l'Algérie,* Algiers, 1922.

Maire, René, Dr., *Un Voyage botanique dans le Sahara Central.*

Rolland: "Sur les grandes dunes du Sahara," *Bulletin Soc. Geol. Fr.,* Vol. X, 1882.

Schirmer, H., *Le Sahara,* Paris, 1893.

Walther, J., *Das Gesetz der Wüstenbildung,* Berlin, 1900.

PART II

PHYSICAL EXISTENCE OF THE SAHARA
PRESENT AND PAST

IV
Structural Formation

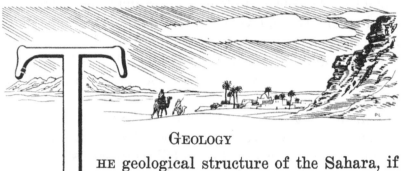

GEOLOGY

THE geological structure of the Sahara, if we attempt to describe it only along general lines, is really very simple. Along the border to the northwest lies the Atlas range, a young folded range comparable to the Alps, and which, moreover, does form a part of the Alpine system. But although on its southern declivity this mountain system is steppelike, even desertlike in character, it serves as a frame to the Sahara, rather than forming a part of it.

The true Sahara, in all the remainder of its extent, is just the opposite of a young folded range. It is more nearly an equivalent, if we must make a comparison, to the central plateau of France; certainly not to the Alps or the Pyrenees. It might even more justly be compared to the Russian, Siberian or Canadian platforms, being what geologists call a "shield" ["basement complex" or "oldland"]: a block of the earth's crust that has remained in fixed position for countless ages.

In certain parts, particularly in the Algerian Sahara, we do find traces of very ancient folded ridges, contemporaneous with, or even older than, those which are found in the French Central Massif or in the Rhenish Schist Massif and designated as Hercynian by the European classification. But in the Sahara as in Europe the old worn ridge, leveled to the roots of its folds, has in the course of time long since disappeared as a range. Only remnants now remain, forming what is known as a peneplane, but having here the appearance of a plateau.

Throughout the remainder of the Sahara there are immense areas where the Carboniferous limestones, Devonian sandstones, and even Silurian sandstones appear in beds that are almost horizontal. It is very rare to find deposits as old as these in horizontal beds, just as they were laid down. They are seldom encountered anywhere on the face of the earth except in places where there are other such platforms, as for instance on the Russian Oldland or on the Canadian Shield. These very ancient primary rocks form the foundation of the whole Sahara.

There are immense stretches, however, comprising perhaps half the Sahara, where this substratum or foundation disappears under an overlay of more recent rocks. These include the Cretaceous limestones of the Algerian and Tunisian Southern Territories, of Tripolitania and of Cyrenaica; the Nubian sandstones of the Libyan Desert, also of Cretaceous age; and the Miocene limestones of Marmarica. Enormous areas are covered uniformly by these secondary or tertiary sandstones and limestones, because they, too, are laid down horizontally; although in many places, particularly in the Algerian and Tunisian Sahara, one may detect beneath this overlay the fibers of the ancient peneplane, as one may sense a skeleton beneath the skin that covers it.

Crossing this complexity we find recent eruptive

THE "REG"

Photograph by Gautier

rocks that have made their appearance in great abundance. The most outstanding of the mountainous massifs of the Sahara, the Tibesti, the Aïr and the Ahaggar, are all volcanic. Anything, indeed, that rises sharply above the general level of the desert platform is more than likely to be of volcanic nature.

OROGRAPHY

While the desert is indeed a platform, with the plateaus and plains by far the most predominant forms, there are nevertheless very considerable differences of altitude in its entirety. In the Tunisian Southern Territories and along the Egyptian and Tripolitan frontiers there are certain fairly extensive areas which lie from 50 to 100 feet below sea level. In contrast to this, we have Emi Kusi in the Tibesti rising to a height of more than 11,000 feet and Mount Ilaman in the Ahaggar with an altitude of almost 9,850 feet. These, to be sure, are volcanoes; but the very abundance of volcanic formations in the area is in itself clear evidence that the Saharan platform is crossed by lines of fractures which have recently been active and are probably still active. Thus there are also many other places where blocks of the peneplane itself, upraised along fault lines, have been rejuvenated by erosion and now present the appearance of steep cliffs, of ranges, or of fragments of ranges.

This Saharan relief, which at first glance appears confused, is really, on the contrary, governed by a law which is quite evident and very simple. The series of volcanoes aligns itself roughly but plainly in an east-west direction across the Central Sahara between the Tibesti and In Ziza, passing through the Aïr and the Ahaggar. This is also the direction followed by the

Atlas, and by the south coast of the Mediterranean in its general trend. Taking this same direction across nearly the whole of the northern Sahara there is likewise a vast system of depressions which may be followed almost without interruption from Cairo to In Salah in the extreme south of the Algerian Military Territories. The northern sides of these depressions are almost everywhere marked by southern-facing cliffs which also constitute one extensive system, from those which form the southern boundaries of Marmarica and Cyrenaica, to the Hammada Homra in Tripolitania, the Tinghert which continues it into Algerian territory, and the terminal cliffs of Tademaït. At the foot of these cliffs lies a series of oases, including the famous ones of Siwah (Jupiter Ammon), Jerabub and Aujila, as well as those of Tidekelt which are found much farther along after an interruption. This chain of oases also aligns itself in the same general east-west direction—which is that of the latitude.

There is another trend of equal importance which makes an approximate right angle with the first. This is roughly north-south, oriented more or less with the longitude. It is the direction of the trough of the Red Sea, of the Arabian range, and of the Nile Valley beginning at Khartum. Geologists have found this longitudinal direction repeated in two great parallel faults striking across Tripolitania, with the notch of the Gulf of Sidra a prolongation of the same system. And finally, it is this longitudinal trend that is of predominating importance in the Algerian Southern Territories.

A bathymetric chart of the Mediterranean Sea shows this same checkerboard pattern of two trends, approximately orthogonal north-south and east-west;

while in the outline of its coasts the design is very apparent. All of which would make it seem as if, in this extensive portion of the earth's crust, there had been something like a tendency toward "geoidal torsion." To this circumstance, the relief of the Sahara as a whole, considered on a large scale, owes a grandiose simplicity of design.

V
Fundamental Laws of the Desert Surface Relief

To those of us who have lived our lives in temperate climates and in countries that are normally drained, the desert is a world apart. Its landscapes would be as strange to our eyes as those of the polar zone could be. Our vision, our geographical concepts, our vocabularies, even our imaginations, have been formed under totally different circumstances. It would be impossible to present a clear picture of the Sahara without first giving a general idea of the laws which govern the desert surface relief.

The Forces Involved

There is often a tendency to exaggerate the influence of eolian erosion in the formation of the characteristic relief of the desert. This perhaps is natural, for the wind of the desert is the only element of life and movement in a domain of immobility and death. Any journey in the desert is a constant battle against this sand-laden wind, sometimes in critical moments a physically painful one. Moreover, in the details of the landscape the eye encounters in every direction the evident scars of eolian erosion. Consequently the traveler's first impression is that he has penetrated into a realm where the wind reigns supreme, and he is very apt to over-estimate its power.

When the Suez Canal was first projected, the adver-saries of the scheme objected to it on the ground that the torrents of sand carried by the desert wind would be sure to choke the canal. Yet, in actual experience, the canal company finds it necessary to devote only a small portion of its annual budget for dredging; which fact in itself shows that our imaginations are unequal to the task of measuring the actual effects of the wind. We must therefore consider carefully what other forces may be at work, before we form any hasty conclusions.

We must particularly keep in mind the very sig-nificant fact that there is no region on earth, even in the Sahara, where rain never falls. And on ground unprovided with vegetation, where the extremes of temperature splinter the surfaces of the rocks and reduce the clay substances to dust, the rare but ex-tremely violent storms leave ravages of extraordinary erosion, so that such fluvial action as there is must be out of all proportion to the actual amount of water involved. Also, generally speaking, running water freighted with earth is necessarily a more powerful instrument of erosion than an air current, however charged with sand the latter may be. And in addition, the action of a stream guided by its bed is concentrated along a straight line, while that of the wind is dissi-pated over a very extensive territory. The fact is that in the formation of the desert surface relief the forces of water and wind collaborate; and in the present state of our knowledge it is not always easy to determine exactly in detail the proportions of each, even in par-ticular cases.

Orogenic Action: The Closed Basins

The most outstanding characteristic of desert topography is the prevalence of closed basins, where the general slopes, instead of inclining toward the sea, converge toward the center of a bowl. The Nile is the only great river of the Sahara; and, aside from unimportant coastal streams, it is the only one which rises in the mountains and empties into the sea. Such others as there are, are "wadis" or desert streams, terminating in an alluvial region in a closed continental basin. The Sahara as a whole is an intricacy of these closed basins. Their depressions vary considerably in altitude, and the centers of some of them lie well below sea level: the famous oasis of Siwah on the borderland between Egypt and Tripolitania is −65 feet, and the Great Shotts of southern Constantine at Melghir are nearly 100 feet below the level of the sea. This prevalence of closed basins is a structural trait common to all deserts.

The hollows of desert topography have sometimes been attributed to the predominant influence of wind erosion; and it is quite possible that in certain cases some specific basin does owe its origin to deflation: that is, a hollowing-out of an outcropping of soft rock. But as a general and comprehensive explanation, the eolian theory alone is certainly not adequate to account for the presence of these vast depressions.

All over the surface of the earth the orogenic movements of the terrestrial crust have a tendency to create topographical hollows. But in regions of plentiful rainfall and normal drainage, the incessant labor of the rivers fills up these basins with sedimentary deposits, at the same time wearing down their edges by erosion.

thus maintaining the general slope from the mountains to the sea. In such countries the fluvial erosion reestablishes the slope faster than the orogenic deformations destroy it. In a desert there is an inverse tendency which upsets this equilibrium; the rivers supplied by scanty rainfall have not sufficient erosive power to counteract the orogenic action.

At the same time, there is probably not a single desert basin where the effect of wind erosion must not be considered an important factor. The lowest point of a closed basin is naturally the place where the alluvial deposits accumulate, and the desert wind exercises all its powers on these light sands and earth; so that while the rivers have a tendency to fill up these topographical hollows, the wind opposes the effect of sedimentation, diminishing or even destroying it. Thus, if it has not actually contributed to the formation of the basin, it does help to maintain or even to deepen it.

We must therefore not attempt to account for the closed basins of the desert by mathematical deduction from a single principle. While giving all due credit to the importance of the part played by the wind, we must be careful to keep in mind the important and in many cases decisive part played by fluvial erosion, even though in this case it would seem to be a purely negative one, consisting in its failure to destroy the hollows once they have been created.

Laws of Fluvial Erosion in the Desert: The Desert Peneplane

Fluvial action in desert countries does not obey the same laws as it does in countries of normal drainage, either in the process of sedimentation or in that of erosion. For it makes a most profound difference whether a watercourse, supplied by plentiful rains, is going to empty into the sea, whose capacity for receiving it is practically unlimited, or, fed by infrequent storms, will terminate in a continental basin.

A normally drained country is in all parts dissected from the center to the circumference by great powerful rivers which have a common base level, the sea. Moreover, this base level is practically immutable, and the mouth through which the normal river discharges remains relatively fixed. All these rivers then tend to deepen their beds to the ideal term of the coastal zero throughout the entire length of their valleys; they have the power to rip open the ground and cause a general and uniform peneplanation, which the wadis, within closed basins in the continental interiors, do not have. Likewise, the sands, pebbles and alluvial sediment which are carried along relentlessly by the stream must invariably, after more or less frequent shiftings and after a time more or less protracted, end by being emptied into the sea.

When the desert wadi terminates in an alluvial region, the sediment which it carries remains in the basin; it comes to rest there, heaps up and becomes stratified. It has not left the continent, but has simply changed its place on the surface. Now these depressions, as we have said, are at various and sometimes considerable altitudes; the wadi, moreover, has for its base level the terminal accumulation of its own alluvial deposits, a level which it is itself constantly raising. The sedimentary region which serves it as a mouth is consequently always uncertain and unstable. Each terminal outlet, by the very fact that it is a terminal outlet, tends naturally to become obstructed by the accumulation of detritus and the raising of its level.

Each sedimentary region is a potential delta, and the wadi is continually forced to seek new courses. The alluvial surface thus extends itself indefinitely, presently lining and filling the whole interior of the basin, which is usually of enormous extent. In this way is formed one of the characteristic features of the desert landscape, the plain which appears to the eye in every direction as level as the sea. This is the plain which the Arabs in the Occidental Sahara call the *reg,* and in the Oriental Sahara the *serir.* [See illustration, p. 30.]

This overlay of the regs in turn acts as a protection for the depressions against fluvial erosion, so that the original structural formations are preserved. The wadis have thus been building up a barrier against themselves. In a desert as arid as the present Sahara a great number of the rivers are short and intermittent torrents, with an alluvial region beginning immediately on their emergence from their native mountains, so that their erosive power is lost almost at once. The whole base of the mountain is protected against erosion by a continuous belt of alluvial cones, prolonged to the very horizon and beyond.

The mountain itself, however, is attacked by the erosion with a force which all the climatic influences tend to increase. On its bare flanks, which have no vegetation to protect them, the denuded rocks, exposed in the moistureless atmosphere to extreme and abrupt variations of temperature, burst into splinters or disintegrate into impalpable dust; or they are shattered by the shock of the storms and break off in great slabs with the undermining of the swelling torrents.

The effect which this produces has no parallel in our climates except that which has been obtained by glacial action on the summits of the Alps. Forms indeed are

often found in the Sahara which recall the Alpine pin-
nacles, rocky peaks which are as difficult of ascent as
are those of the Alps. Such broken stumps, abrupt and
jagged, are all the more striking because they here
emerge without any transition from the infinitely flat
reg, creating a landscape whose appearance is that of a
rocky archipelago dotting the surface of a calm sea,
as the Cyclades rise out of the Ægean.

Mount Ilaman, the highest peak of the Ahaggar, is a
pinnacle of this nature and has a most impressive aspect; while to the north of the Ahaggar rises the famous
Garet el Jenun, the Mountain of the Jinns, so called
by the natives because its summit has never been
touched by human foot and seems to them to be reserved
for the jinns, or spirits. In exactly the same way the
extinct volcano of In Ziza, jutting from the immensity
of the surrounding reg, makes an abrupt and almost
ludicrous contrast; while the mountains of Aïr seem
to be poised on the plain like cones of sugar upended
on a table.

In order to bring out this aspect of the country, so
strange to European or American eyes, the Algerian
geologists have sought for various comparisons. One
evokes the prow of a ship springing up out of the sea.
Another, describing the half-buried fragments of a
range, calls it a procession of caterpillars traveling
head-to-tail along a road. It is one of the most characteristic features of the desert surface relief.

All erosion, working on the relief of a continent,
tends to reduce it to a peneplane; and it is probable
that the tendency of the desert erosion is to produce a
peneplane of a very particular type, whose characteristic trait would be exactly the persistence of such
abrupt and isolated stumps of bare rock, scattered ap-

Photograph by Désiré

THE PEAK OF MT. ILAMAN, SUMMIT OF THE AHAGGAR

parently at random over a more or less uniform tableland. It would seem that fluvial erosion, such as occurs
in a desert climate, must certainly be the essential factor in this plan of relief. [See illustrations, pp. 40, 50,
62, 222.]

EOLIAN EROSION: THE DESERT LANDSCAPE

However considerable the part played by fluvial
erosion in desert country, the effects of wind erosion
must naturally also be immense. They leap to the eyes
in a host of minor exterior details of the relief, such as
polished boulders, isolated stone pillars thinned at the
base, pierced rocks and rocky walls riddled with caves
and sculptured in fantastic forms. For the wind has
sought out every least difference of density in the rock,
every interstice between rounded boulders, every crack
or crevice into which it could insinuate itself and pro-
ceed to excavate within. Even the beds of pure gravel
which cover to varying thicknesses such immense expanses of the reg or serir, are evidences of the wind's
craft, for they are almost free of any admixture of
foreign substances and have the consistency of a garden
walk.

One very particular aspect of the desert surface is
what the Saharan nomads call the "hammada," a table
of denuded rock with an appearance of having been
dusted and varnished. These hammadas roughly resemble a shelf extending indefinitely to the limits of the
horizon and well beyond, for in the Algerian and Tripolitan Sahara one may travel along them for days together. They are great limestone or sandstone plateaus
whose rocky top layer has been despoiled of its covering
of light earth and laid bare by the eternal sweeping of
the wind. Such are the superficial results of eolian ac-

tion, those which affect only the epidermis or outer skin of the desert; although assuredly in the course of ages they must have incalculably profound effects.

The true domain of the wind, however, would seem at first glance to be the erg. The term *erg* is applied by the Saharans to the vast sandy expanses which are occupied by the great masses of the dunes, and which extend over enormous areas. The Libyan Erg, which is probably the largest on the surface of the earth, is the size of France; while the two great ergs of the Algerian Sahara are each some 200 miles in greatest diameter by nearly 100 miles in width. These latter, known as the occidental and oriental ergs, are apparently the best known, or at any rate have been the most thoroughly studied. Now these vast seas of sand with their mighty tumbled waves are obviously at the mercy of the wind, while fluvial erosion has absolutely no direct action upon them; yet even in this case, relatively so clear-cut, there are indirect effects of fluvial erosion which are of very great importance.

Attempts have sometimes been made to explain even the existence of the sand itself by wind erosion alone. The theory is that these great masses, so enormous as to confound the imagination, are made up of sand which has been detached, grain by grain, through corrasion from the parent rock, particularly from the beds of sandstone. But while the effects of corrasion are obviously quite real, the fact is that corrasion encounters certain obstacles which check or greatly diminish its effects. One of these is the desert patina which is so frequently found to cover the hard rock, a very striking phenomenon and one which has been scientifically studied, especially by Walther in Egypt. It is a crust of chemical substances exuded by the porous rock, brought to the surface by capillarity and fixed by evaporation. Oxide of iron colors it dark red or black; and in places where it has splintered off, the lighter heart of the rock shows in vivid contrast. It is this patina which gives the shiny, varnished appearance not only to the isolated rocks but to the whole immense extent of the hammadas; its somber coloring is reflected in the name of the great Tripolitan plateau, the Hammada Homra or Red Hammada.

This crust is very hard and constitutes a definite obstacle to corrasion; it is even possible for us to formulate some measure of its resistance. On the sandstones in the Algerian Sahara are to be found quite a large number of very ancient engravings, some of which are approximately dated by their very subjects, as for instance those which represent Ammon Ra, the god of Thebes. These carvings, although exposed on the walls of bare rock to every violence of the wind for several thousands of years, seem now as fresh as the day they were made. We also have evidence that the desert climate preserves indefinitely the most delicate bas-reliefs in stone, as witnessed by the architecture and sculpture of Egypt; in fact, the obelisk of Luxor has deteriorated more in fifty years in the Place de la Concorde than it did in fifty centuries on the banks of the Nile.

Yet we must remember how speedily these same hard rocks, which offer such a long resistance to the wind, can crumble under the shock of the Saharan storms, shattering into splinters and slabs, while the debris is rolled and ground as it is swept along in the bed of the torrent. The great Saharan basins, which as we have shown are alluvial regions, are carpeted to immeasurable thicknesses by light sand whose fluvial origin is indisputable. Now it is in these alluvial basins that, as a general rule, the ergs are formed; and these sedimentary regions seem to be the places particularly favored by the big dunes.

It is often observed that the Saharan dunes show differences of color; some of them are white and others are golden. The latter are the great dunes, ancient and mighty, which for ages have been exposed to the action of the wind; each separate grain of sand has had time to oxidize, to redden, from its contact with the air. The white dunes are usually the smaller sand waves around the edges; and the natural supposition is that they have just come into being, that their grains have not yet taken on the desert patina and still show the original color of the alluvial sand. It is impossible to escape from the idea that in all this there is, in some measure, a probable sequence of cause and effect. Everything tends to indicate that it is fluvial erosion, and not wind erosion, which has furnished the erg and the dune with the greater part of their constituent material of light, free sand.

But the sand furnished by the alluvium is more or less mixed with mud, and the sand of the dunes is a pure sand. This is because it is subjected to a ceaseless winnowing by the wind, which blows out all the lighter particles, leaving behind only the heavier residue. Of all the manifestations of eolian action, this winnowing process is perhaps the most unique, the most direct and far-reaching in result. This is natural, since the wind is working here with the most tenuous of substances, with those which of all substances offer it the least resistance.

The process itself can be directly observed in the atmosphere, which, throughout the entire Sahara, is only partially transparent at any time of the year, as can be testified by anyone who has taken an hour-angle observation. Even with the clearest sky a dark-colored glass is useless to take a sun sight with sextant or theodolite; only the most lightly tinted can be used. Apparently the desert air is eternally freighted with dust in suspension. Pocket watches also bear witness to this fact, for those which are not hermetically sealed will become so thoroughly clogged with dirt as to stop within a week.

In the French Sahara, especially to the south in the vicinity of the Sudanese steppes, traversed as they are by active tropical rivers, the blasts of wind are observed to be accompanied by a darkening of the atmosphere, which deepens to a black opacity, creating night in full midday. Chudeau has portrayed these soot storms which, with plumy crests and mushroomed whirlwinds, can be seen approaching from the distance on the rim of the horizon. Analogous phenomena were also observed by the Tilho expedition in the northeastern part of the Chad region. The Egyptian khamsin is also an opaque wind, whose ingredients are doubtless drawn from the mud flats of the Nile.

The quantity of dust thus floating in the atmosphere at all heights must be enormous. The particles are so tenuous and so light that they cannot fall by themselves and must remain indefinitely in suspension until finally the air currents carry them beyond the desert zone and into regions of normal rainfall. There at last they are brought back to earth by the rain's periodic washing of the atmosphere.

The accumulation of these desert dusts has in the course of ages brought about a very distinctive and well-recognized geological formation which is unevenly distributed around the fringes of desert zones. A good deal of attention has been directed to this formation for some time, especially in our temperate zones and throughout the northern parts of the Old and New Worlds. It is known as the loess and is now admitted by geologists to have been produced in this way. It is in exactly the same manner that the slimy substances in the oceans are carried far from the coasts and distributed at large over the great deeps.

The sand of the erg thus winnowed, the wind now exercises upon it an action of stirring up and transferring. It is this action which is the more apparent at first glance, and to it have been attributed all the surface formations assumed by the sand, the characteristic heaping-up of it into dunes. It has often been studied, though this does not signify that anything is definitely known about it as yet. All we know for a certainty is that the dune does tend to assume an asymmetrical form which would seem to be related to the wind, since it is characterized by a long and gentle slope on the side directly exposed, while the sheltered side falls away abruptly in a tottering wall.

Those who specialize in the study of the dune, however, often go so far as to claim to have determined the elementary desert form, in which the erg would consist of a composite of small circular dunes, arranged in curving lines and advancing in eternal progression across an even surface. The unit form indicated by this theory would be the small crescentic dune which in Turkistan and Mongolia is called *barkhan;* and this would indeed be the elementary grouping of sand in motion when the sand had a tendency to group by itself. But in the Sahara, perhaps because of the extreme dryness of the atmosphere, the sand in motion exhibits no

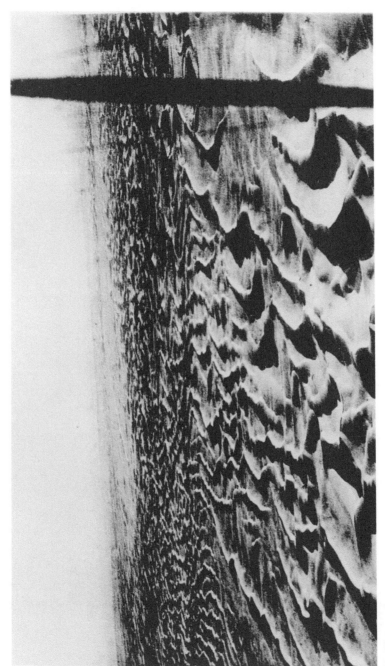

Air Photograph

THE DUNES FROM THE AIR : IN THE REGION OF EL WAD

tendency toward cohesion but tends to retain a kind of individual independence of the grains. Moreover it must be noted that in the Sahara the typical barkhan is very rare.

A characteristic sign of its rarity is the fact that the vocabulary of the indigenous Saharan, always so rich and expressive of exact shades of meaning, does not have a single word which corresponds to the term barkhan. A few examples of the formation were found by Chudeau in Mauretania, not far from the Atlantic Ocean, and Nachtigall has shown some very clear ones in the regions to the north of Lake Chad. But one must search carefully through the bibliographies of Saharan literature to find any reference to this formation, and the barkhan theory would never have been formulated from observations made in the Sahara alone. We should not, therefore, be too hasty in adopting the barkhan or wind-formed dune as the elementary unit of the ergs. [See illustrations, pp. 12, 46.]

Although there are places where the sand takes the form of moving waves—as for instance in the region of the Lower Tuat where these waves advance from the east, entering the palm groves of the oasis at one end and passing out at the other—the Saharan dune customarily appears in enormous stationary masses localized in the great ergs, and there is no perceptible progression. These dunes assume certain formations which constitute a definite plan of relief for the erg, and the indigenous vocabulary permits us to analyze the features of this relief.

The *sifs,* that is to say, "swords," are long ridges, slightly incurving in the form of yataghans, with an edged crest and steep slipping sides, practically insurmountable. This formation is the only one which obviously might bear some distant relation to the barkhan. The *oghurds* are massive mountainous peaks, rising considerably above the general level; from their summits the whole erg lies spread out at one's feet. The *fejj* or *gassi* constitute a more curious feature, for the terms, which mean literally "necks" or "closed grounds," are used to designate certain very long, sand-free passages which in some cases traverse the entire erg from end to end, or at least by connecting with one another greatly facilitate the crossing of it. [See illustration, p. 176.]

Now within the limits of human experience these sifs, oghurds and gassi have formed perfectly definite features of the topography. To the native guides they are landmarks which are recognized unhesitatingly. The French officers of the meharists, or camel corps, have been in intimate contact with the ergs of the Algerian Sahara for more than half a century, and they have noticed no change in them. The post at Taghit is situated at the very foot of an oghurd, and in thirty years the distance between the wall of the station and the edge of the oghurd, which is something like ten yards, has not perceptibly changed. Moreover, the palm groves of Taghit, which are several centuries old, are not bothered in the least by the proximity of the dune; the natives have not even any idea that such proximity might be dangerous. In fact, the very bulk of the oghurd serves as a protection for Taghit, since other palm groves in more open territory do stand in some danger of being choked with sand. Even this, however, does not dismay the natives, for they have traditional ways of protecting their groves, such as hedges of palm leaves fixed in the ground, which make an effective defense against the onslaught of the drifting sand. [See illustration, p. 88.] But there is no known case of a palm grove having been destroyed by the advance of the dune itself.

It is unfortunate that no attempt has ever been made, either in the Sahara or elsewhere, to compile a topographical map of the erg; for such a map, depicting these various features, should be of great aid in determining the law which governs the formation of the dunes. As it is, we may at least venture to suggest the general reason for their stability: in order to bring about the beginning of a dune, there must be some obstacle which blocks the sand. The dune is determined by some structural feature, and is thus dependent upon the underlying surface relief, not only for its form but for its fixity and permanence. It is apparent that the oghurds have a rocky skeleton; while the gassi are invariably plains without relief, which accounts for their being free of sand.

An erg then may be defined as a mantle of sand which masks a relief deeply engraved in the rock. If we had a good map of the erg we should be able to see this relief showing up through the covering of sand. Naturally this underlying relief must have been in large measure the work of fluvial erosion, so that in this case as in everything else pertaining to the desert we find the close collaboration between the forces of wind and water.

There is, however, one particular erg in the Libyan Desert which is of a very distinctive nature. This is the Abu-Moharique, in the vicinity of the Nile. It is a chain of dunes, rectilinear and continuous over five degrees of latitude, yet nowhere exceeding a few miles in width. No entirely satisfactory and acceptable explanation of this erg has been offered, with which all Egyptian geologists are in accord. It seems to have no clear, incontrovertible connection with the existing surface relief; and in this one case a purely eolian explanation seems unavoidable. Can it perhaps be that Abu-Moharique marks the boundary between two zones ruled over by different winds, a kind of atmospheric frontier along which the sand accumulates? In any case, it is one small detail illustrating the variety of features and widely diverse manifestations of the desert influences to be found in the Sahara.

It must be clearly understood, of course, that it is the dune itself which is fixed, and not the grains of sand which compose it; for these are in almost constant motion. When the wind is blowing a gale, as it frequently does, the "smoking dune" becomes a magnificent spectacle, wreathing itself in plumes, with a haze of sand obscuring the horizon. The dune itself retains its place and its general outlines only because new grains of sand have replaced the ones that have been blown away. We have already mentioned the moving sand waves, which constitute a real menace, particularly in the oasis of Lower Tuat, where they sweep the groves from one end to the other.

Now a good share of this mobile sand, eternally shifting, must finally come to the ends of the desert, and so eventually be lost in the Atlantic or the Mediterranean, just as are the argillaceous dusts, which are carried so far to build up the loess. Consequently, however stable the erg itself may be, it is so only within the limits of human memory and observation. If we consider it in the light of geological chronology, every erg as a whole must tend toward displacement in the direction of the predominant wind. Thus deflation must after all, in the course of time, exercise quite a consid-

Photograph by Désiré

THE TRIDENT OF KUDIA, IN THE AHAGGAR

erable effect upon the alluvial filling which lines the closed basins.

Nevertheless, we shall presently see how stubbornly these sedimentary deposits have persisted throughout geological ages. This is because eolian erosion, by its own efforts, is continually building up obstacles against itself. One of these is the *reg,* which extends over enormous expanses, particularly in the French Sahara, where perhaps half the total area is covered by this type of surface. In the case of the reg, all the lighter particles of the superficial layers, not only the clay but the sand as well, have been scattered far and wide by the activity of the wind. But the heavier gravel has been left behind; and this bed of gravel, which is the characteristic feature of the reg, protects anything sealed beneath it, constituting a definite obstacle to further deflation, and one whose strength increases with its depth—which is to say, of course, with time.

In other cases there is a different process which enables the mobile sedimentary deposits to protect themselves directly against eolian erosion in much the same way as we found the sandstone surfaces to protect themselves by means of a hard patina. This is what Algerian geologists call the "croûte calcaire," and is the same as the "caliche" of the Americans. It occurs wherever there is a deep-lying sheet of water below the alluvial bed, so that with the surface constantly swept by a dry brisk wind, the capillarity, enhanced by an intense evaporation, leads to the deposit of a calcareous crust which becomes very hard and may attain a thickness of several feet. It is particularly well developed on the slopes of the Oran Atlas, where it is known as the sub-Atlic hammada, and where all the conditions to explain its presence are found together.

For certainly as long as the Atlas itself has been in existence, water can never have been lacking within the depths of the strong mass of unconsolidated material into which the alluvial cones from the Atlas here merge. Except at points where the play of a fault, or more frequently the erosion of a wadi, has broken the continuity of the protecting shield, these sedimentary deposits remain permanently sealed beneath this thin hard crust, and absolutely protected against deflation.

At this point it would be well to consider a question which has often been raised: What lapse of time has been required by the desert climate to create the Sahara of today?

BIBLIOGRAPHY

GAUTIER, E.-F., "Sahara Algérien," *Mission au Sahara,* Vol. I, Paris, 1908.

MARTONNE, EMMANUEL DE, *Traité de géographie physique* (Chap. X), Paris, Armand Colin, 1909.

PASSARGE, SIEGFRIED, *Die Kalahari,* Berlin, 1904.

—— "Rumpffläche und Inselberge," *Z. D. G. G.,* Vol. LVI, 1904.

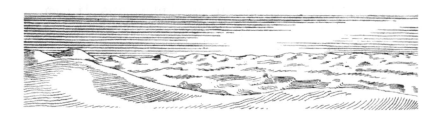

VI
The Geological Past

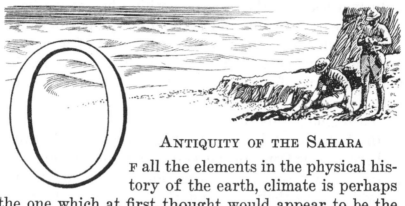

ANTIQUITY OF THE SAHARA

O F all the elements in the physical history of the earth, climate is perhaps the one which at first thought would appear to be the most unstable. The imagination sheers away from the idea of a climate which at any given point on the globe has persisted always the same since the Cretaceous Age or the Silurian Age, or, as it were, since the very beginning. But as we have said before, the distribution of deserts over the planetary surface is after all primarily a function of latitude. And if the pole has not changed its place since the earth's crust solidified (which we do not know for a certainty), it would be natural to suppose a priori that the principal deserts of the world have been, throughout the duration of all the ages, in practically the same location as they are found today.

This supposition has been experimentally verified for the three greatest of the world's deserts, which are the only ones whose geology is well enough known to let us form any conclusions. As regards the Sahara, we now have a good deal of precise and conclusive geological data in support of this contention as to the great antiquity of the desert. The first to be considered are those which indicate the nature and the remote age of the alluvial accretions of the closed basins. In the still unknown heart of the Sahara nothing has as yet been determined concerning the sedimentary deposits which are concealed by the reg; but a great deal of work has been done by the Algerian Geological Service in the northern part of the French Sahara and on the Algerian steppes.

Throughout this region, across the southern slopes of the Saharan Atlas, on the high Algerian plateaus, and even in the Constantine Tell, the ground is found to be incrusted with alluvial accretions containing fossils by which both their age and their continental origin can be established. These deposits are built up, layer on layer, in strata which have been authentically classified from the Oligocene to the Quaternary. Their very arrangement and the enormous areas over which they are spread, as well as their composition and the chemical deposits of salts and gypsum which they contain, all attest that these are beds of alluvium that have been formed within closed basins and in a desert climate, or at least that of the steppe. These indications are particularly marked in the Triassic stage, showing that Algeria at this period was a composite of closed basins, which moreover contained lagoons where both salt and gypsum were deposited in extraordinary quantities.

The sandstones of northern Africa also give us some valuable information regarding the age of the desert. From the Red Sea to the Atlantic Ocean a considerable portion of the surface is masked by sandstone of a uniform and very particular character. It is rather fine grained and dark reddish in color, and contains spheroidal concretions which have caused it to be known in

Algeria among geologists as "sandstone with spheroidal concretions." Now it has been shown by a laboratory test made in Algeria that spheroids of Nubian sandstone from Egypt cannot be distinguished from those originating in the Saharan Atlas. The very diffusion of a formation as regularly uniform as this over such an extensive area would in itself be sufficiently curious. But what is even more remarkable is that while these red sandstones are shown to be a continental formation by the fact that wood is found in them, and sometimes silicified (petrified) trees, nevertheless in certain very localized places perfectly distinguishable marine fossils appear.

From this we perceive that these sandstones, however uniform, were not all formed in the same geological period; they represent not only very remote but also very diverse stages. The Nubian sandstones are Cretaceous, bordering on the Albian stage, as are also those of Algeria, while those of the Sahara are Lower Devonian and Silurian. The problem as to why they should so resemble one another is one for which the geologist Fourtau seems to have indicated a solution. These sandstones are all similar because, in spite of their diverse ages, they have a common origin; they are all solidified, petrified ergs. The Sahara then, at an age as remote as the Silurian, must already have been a desert. We might here note that geologists in studying the North American and the Kalahari deserts are arriving at the same conclusions as to their antiquity.

It must be kept in mind, however, that the stability of climate which all this evidence would indicate is only relative; within certain limits, of course, there have been very wide fluctuations of climate which have brought about temporary changes in the character of the desert, or in portions of it.

THE QUATERNARY SAHARA: THE FOSSIL WADIS

The most apparent traces of a radical change of climate have been left by the period immediately preceding our own. It is the one known to geologists as the Quaternary, and more popularly as the glacial period. To be sure, no glaciers were able to develop in Northern Africa, since it was in too low a latitude; but there as elsewhere the Quaternary climate must have been much more humid than the present one. For there as elsewhere the rivers of today are but puny dwarfs, lost in great valleys which were obviously fashioned for them by gigantic ancestors and which now no longer fit them.

These vast complex valleys, cut deeply into the earth, are the best evidence we have of a previous humid period; they prove conclusively that the whole immense region centering around the Ahaggar massif and extending from the Saharan Atlas to the bend of the Niger was once furrowed by mighty rivers. And while the rivers themselves are practically extinct today, their old networks are still easily distinguishable. This freshness of their forms shows us that the period in which these dead valleys were flowing with running water was one which cannot be placed very far back in the past.

The best examples of these great Quaternary wadi valleys are to be found in the French Sahara, with their principal center of derivation in the Ahaggar massif. Diverging from it toward all points of the horizon, we find these once great watercourses, reduced now to the state of skeletons. Northward the Igharghar can be traced more or less clearly as far as the basin of the

Great Shotts, which is to say the foot of the Tunisian Atlas; while the southward course of the Tafassasset may be followed all the way to the Niger, although recent research seems to have established that this river in reality flowed toward Bilma and Lake Chad.*

The Atlas range, particularly the High Atlas of Morocco, was also a source region for the Quaternary wadis. The most important of these Atlas wadis, as far as we have been able to determine, was the Saura, whose great network is still very clearly defined. The convergence of its arteries can easily be followed to the basins of Gurara and Tuat. In fact there is hardly a spot in this very extensive region where we cannot tell with precision to which of the Quaternary basins it belonged.

We must not, however, be led astray by appearances. The remarkable preservation of the Quaternary fluvial systems, the depth, length and multiplicity of their channels, have given rise to a popular fallacy by which the Quaternary Sahara has sometimes been represented as a country watered by overabundant rains and normally drained, the opposite of a desert. But we know that the mighty Igharghar had for its terminal basin the depression of the Great Shotts southeast of Biskra, a region which has been thoroughly studied, especially as there has long been a project to transform this great bowl, which lies partially below sea level, into an "interior sea." It is today separated from the neighboring Mediterranean only by the Gabes ledge; and on this ledge, in spite of every effort, geologists have been unable to find the least trace of an ancient fluvial junction between the basin and the sea. The very surface relief

* See the Saharan sheet in the *Atlas des Colonies Françaises*, 1934. Prepared under the direction of the author, E.-F. Gautier.

of the depression is that of a closed basin, and its whole vast extent is thickly lined with a coating of sedimentary deposits. Nowhere on its circumference do we find lines of cliffs such as must necessarily be formed on the banks of a lake of fixed level in a country normally drained. Whereas we do know, if only from the example of Lake Chad, that a lake in steppe country, being the terminal alluvial region of a river, has no definite banks. All of which combines to prove that the Igharghar, even in its heyday, when crocodiles disported themselves in its living waters, terminated always in an alluvial region; never did it have force enough to break through the low ledge which separated its terminal basin from the sea.

During their existence, however, these Quaternary rivers did form an open communication route between the Sudan and the Atlas, by means of which the fauna, and doubtless also the vegetation of the tropics were enabled to reach as far north as the Mediterranean. Certain elements of this fauna have survived, not only into our historical period, but even to the present day, which is additional proof that the epoch in which the great wadis flourished was one contiguous to our own.

The *Chromys,* a kind of small tropical fish, is very common at Biskra and in the oases of the Wadi R'ir, which as we have said was the terminal region of the Igharghar. These fish today abound in the water holes and in the irrigation canals of the palm groves; they have even been seen to gush from the wells with the artesian waters, showing that they take refuge where they can, even in the subterranean layers. Recently a much larger fish has been found in the same region. This is the *Clarias lazera,* a Silurus commonly known in English as the catfish. In the earlier days it also was a tropical fish, and while it abounds in Egypt because it follows the Nile, it is really an intruder in the Mediterranean world. Yet in the Algerian Sahara this denizen of the ooze still maintains a precarious existence in the depths of the water holes, and is known the length of the Igharghar from the source to the terminal region.

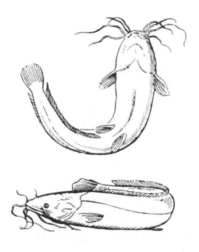

From Duveyrier, Les Tuaregs de Nord

CLARIAS LAZERA (CATFISH)

The catfish and the chromys in the vicinity of Biskra have a companion more celebrated than themselves. This is Cleopatra's asp, the serpent of the charmers, and is the same as the cobra of India. It is likewise an immigrant from the tropics, and its presence in southern Algeria is inexplicable unless we interpose the medium of the Quaternary Igharghar to have brought it there. But the most amazing example is the crocodile. One of these was actually found in the water holes of the Wadi Mihero, an artery of the Igharghar system. This one was perhaps the last survivor; and though it must be considered a biological miracle to find even one such animal in such a spot, it is nevertheless an undeniable fact.

All this seems to put us in very close touch with a period when the Igharghar and the Tafassasset, in contact at their sources, established a communication route of living waters between the tropics and the Mediterranean world. Nor was this aquatic life the only type to take advantage of the facilities thus offered. The Atlas became the home of a residual fauna which has been designated by paleontologists as the Zambezi fauna, some members of which not only survived the Quaternary period but remained in existence even into our own early historical times. In fact, the well-known Carthaginian elephant was the last relic of this fauna, and was not exterminated until the depredations of the Roman ivory hunters.

Here again we must be careful to guard against a rather natural delusion. With such striking biological facts before us, it might seem that this free communication between the Sudan and the Mediterranean had been very much closer to our own period than it really was. But we must remember that it is geological chronology with which we are dealing, not the human and historical measurement of time.

The Carthaginian elephant, made familiar to us by the ancient historians, was a much smaller and less powerful animal than the Asiatic or Hindoo elephant used by the troops of Seleucus and Antiochus; there was no question of this when they were matched one day in battle. But even the elephant of India is less powerful than its congener of tropical Africa. The elephant of Carthage then had already become a distinct species, degenerated and tending toward dwarfism, as would be natural in an animal belonging to a residual fauna. And for an animal species to become thus individualized requires a period of time infinitely exceeding the limits of human memory and experience, even though it be short in geological chronology.

The facts which we have established then are these: that the Sahara appears to have been a desert during very remote and diverse geological eras, but in the Quaternary age, which was the geological period immediately preceding our own, a sharp change of climate in respect to humidity was experienced in the Sahara as well as in Europe and other parts of the world; that during this period certain portions of the Sahara were furrowed by mighty rivers, and for the desert was temporarily substituted the steppe, thus opening to the tropical fauna a route to the Mediterranean; but that these rivers were never powerful enough to reach the sea and thus establish normal drainage.

The Pre-Quaternary Sahara: The Libyan Desert

It is certain, however, that not all of the Sahara was covered by this Quaternary steppe. Judging by the surface relief, there were immense Saharan regions that remained outside the influence of fluvial erosion even during the Quaternary period.

The Libyan Desert, at least the eastern portion of it, has been as thoroughly studied as the Algerian Sahara, for the Geological Survey of Cairo has done admirable work in this locality, including even topographic work. We find that the left bank of the Nile, between the river and the edge of the Libyan erg, shows no fossil system of Quaternary wadis whatever. Nor is it possible in any way to relate the erosion cliffs found here with fluvial erosion by Quaternary wadis. A great rectilinear cliff skirts the Mediterranean from the root

of the Nile delta to the oasis of Siwah, separating Marmarica from the true Libyan Desert; while a more or
less irregular chain of cliffs outlines each of the groups
of Libyan oases: Kharga, Dakhla, Farafra and Baharia. A glance at the map shows us the general plan of
these cliffs on the surface of the Libyan Desert, and its
incoherence is apparent. They could not have been made
by Quaternary fluvial erosion.

Nor does the stony desert of Libyan Egypt exactly
resemble that of the rest of the Sahara. The *serir* of the
Egyptian Libyan Desert is very different in character
from the reg of the Algerian Sahara to which it corresponds. The serir is made up of rounded pebbles,
fairly large and uncertainly balanced one on another,
forming a covering of varying thickness over the surface of the ground. A forced march over such terrain
is torture for man and for his mount; while the English
motor corps, during the campaign of 1916-17, declared
it to be death to pneumatic tires. The reg, on the other
hand, is an immense gravel bed, the choicest terrain for
travel by foot, horse or vehicle.

Now the great regs of the Algerian Sahara, so remarkable for their vast extent, are manifestly the work
of the big fossil wadis which in their lower reaches were
filling their terminal basins with alluvium at the same
time that in their upper courses they were excavating
their canyons. But in the Libyan Desert there are no
fossil wadi valleys with which to connect such alluvial
beds. Nevertheless, both reg and serir are analogous formations, each being a bed of rolled stones left in place
by deflation. The difference is that the serir appears to
be a much more worn formation, as if the deflation had
made use of an infinitely longer period of time to scour
it, to break it up and dismember it.

IN THE LIBYAN DESERT, VICINITY OF KHARGA

Here surely the question of age arises. The reg is geologically young and belongs to the Quaternary stage; while the Egyptian serir, although a formation of analogous origin, dates back to some previous fluvial period so much more remote that all traces of its erosion have been obliterated. It is Pliocene or pre-Pliocene, and some geologists have even gone so far as to consider it Oligocene; in any case, it is a very ancient, pre-Quaternary reg.

Thus the surface relief of the Libyan Desert of Egypt makes an extraordinary contrast with that of the Algerian Sahara. On the latter, the humid Quaternary has left the most apparent of traces; on the former none at all. Should we conclude from this that the humidity of the Quaternary epoch was confined to the occidental portion of the desert? This idea is refuted by the fact that the Arabian range on the right bank of the Nile is carved by dry valleys which are obviously the work of recent fluvial erosion. What we may safely say is this: that fresh traces of the Quaternary rains are to be found only in the highland regions of the Sahara—the Saharan Atlas, the Ahaggar, Tibesti and Arabian ranges; but the beneficent effects of the steppe climate were certainly not felt throughout the entirety of the desert.

Thus we are enabled with a certain degree of precision to reconstruct in our minds at least one of these deserts which have succeeded one another in the northern portion of Africa since the Silurian age. We have a fairly clear picture of the penultimate or Quaternary Sahara, a desert very different in character from the Sahara of our own age. And since here as elsewhere the present is greatly dependent on the past, an understanding of this Quaternary Sahara will be of great value in helping us to an understanding of the Sahara of the present day.

BIBLIOGRAPHY

FLAMAND, G. B. M., *Recherches sur le haut pays de l'Oranie,* Lyon, 1911.

FOURTAU, R., "Sur les grès nubien," *Comptes Rendus de l'Académie des Sciences,* Nov. 10, 1902.

GAUTIER, E.-F., *Structure de l'Algérie,* Paris, 1922.

VII

Superficial Circulation of the Waters

HERE is hardly any need to emphasize the fact that life in the Sahara is dependent on its resources of water, and that in consequence the existing rivers are of primary importance. The ones which have the greatest value are naturally those which originate outside of the desert, in regions better watered. The general level of the Sahara itself is relatively low, while not only is it dominated to the northwest by the powerful Atlas range, but to the south it is also encircled by the mighty massifs of Central Africa: the Futa-Jallon, Adamawa and Abyssinian ranges. Thus from the south as well as from the north the general slope of the terrain directs toward the Saharan depressions the waters of rains which have fallen beyond and sometimes very far from the desert zone. This is a sort of hydrographic swindle which enormously improves the habitability of the desert to the detriment of the neighboring regions.

THE TROPICAL RIVERS

By far the most important of these contributions are the ones which come from the south. The Central African mountains not only cover extensive areas, but by reason of their latitude they belong definitely to the true equatorial zone. Consequently great quantities of tropical rains are brought to the Sahara by such rivers as the Niger, the Shari, which ends in Lake Chad, and especially the Nile. What the Sahara does with them remains to be seen.

THE NIGER

The Niger is a Saharan river by reason of its loop, on which Timbuktu is situated. Rising in the mountains of Futa-Jallon on the coast of the Gulf of Guinea, it first turns its back on this gulf and flows straight north as far as the Saharan zone. There it reverses its direction and turns again toward the Gulf of Guinea, into which it finally empties. Now the very outline of this loop on the map suggests the idea of what geomorphologists call an "elbow of capture," and the hypothesis thus suggested is fully confirmed by an examination of the terrain.

The two parts of this loop, the Upper Niger and the Lower Niger, were originally two separate rivers whose destinies have been joined by a recent capture, each of them still preserving a distinctive character of its own. In the western half of the loop, between Jenne and Timbuktu, the Upper Niger follows a blind course, spreading out in a lacework of extensive marshes and swamps; and at its flood it strays from its own bed, crosses a low ledge, and overflows into the neighboring and ordinarily independent basin of the Fagibine.

This is a typical desert alluvial region. Formerly it extended much farther to the north and northwest, but, as this is the region of the Juf, one of the most inaccessible and unknown parts of the Sahara, it is impossible to state forthwith precisely what may have been the past relations between the terminal wanderings of the Upper Niger and this great Juf depression. Certain indicative features however are distinctly apparent.

The Juf is a shott region, where strong and regular beds of salt alternate with layers of clay; for centuries the salt mines of the Juf have played an important rôle in the economic life of the Sudan. Trarza was the center of this industry in the Middle Ages, but since the beginning of the seventeenth century it has been Taudeni, which is still in active operation to the present day. Now according to Chudeau the altitude here is only 460 feet above sea level, as against that of Timbuktu, which is 885 feet—a drop in elevation of over 400 feet for a distance of some 375 miles as the crow flies. An arid plain practically lacking in any surface relief stretches between Timbuktu and Taudeni; it is strewn with small dunes, and fresh shells of Nigerian stagnant-water mollusks are found abundantly throughout the terrain. Even more significant is the fact that these same mollusks are still in existence at Taudeni. It is certain then that the ancient alluvial region of the Niger reached at least as far as this; and even now the water of Taudeni, which is sufficiently abundant to hinder the exploitation of the salt mines, can come only from the Niger, distant as it is, by subterranean infiltration.

This Upper Niger with its clearly defined alluvial region is in striking contrast with the behavior of the Lower Niger. The change takes place abruptly, without transition, where the river passes through the gorges of Burem (or more exactly Tosaye), a shallow groove which it has cut into a transversal ridge of ancient primary rocks. Above these gorges the river is still the Upper Niger of the uncertain course and vast inundation flats. Below, it suddenly becomes the Lower

Niger, its waters flowing swiftly and precipitously through a valley all at once more clearly defined, and in an entirely different direction. It is now headed southward, toward the ocean. Burem is the point where the Lower Niger, with its swift waters and its power of erosion, has captured and drawn off the stagnant waters of the alluvial region.

This is not the only point in the Niger basin where evidence is found of such a capture to the benefit of the ocean and the detriment of the alluvial region. An extremely sharp elbow of capture is indicated in the upper reaches of the Black Volta, which empties into the Gulf of Guinea on the Ashanti-Togo frontier, making it apparent that the upper portion of the Volta was formerly a tributary of the Upper Niger until captured by the Lower Volta to the further impoverishment of the desert.

THE SHARI AND LAKE CHAD

The Shari is almost an exact counterpart of the Niger, except that it terminates in Lake Chad. It has its source in the tropical zone and the western half of its network is fed by streams from the mountains of Adamawa. Therefore it too directs toward the Sahara an abundance of tropical rains.

It ends today in Lake Chad, a great shallow expanse of water, strewn with islands, merging in many places into swamps, and having no definitely defined banks. The lake in fact varies so much in size and shape within a few years, or even from one year to the next, according to the amount of water discharged into it by the Shari, that there is no agreement among the maps of the region drawn up by the various European expeditions which have visited it at different intervals. This then is really the present alluvial region of the river.

The lake certainly has no visible outlet and seems the typical terminal basin where the waters evaporate. But if this were true, they should then leave a residue of salts, and Lake Chad should become brackish. Instead, the water is fresh and drinkable. Travelers are always astonished by this fact, which does seem inexplicable unless we suppose the existence of some secret underground outlet. Chad remains a fresh-water lake only because it is not a true terminal basin; the waters which seem to stagnate here are in reality flowing through it, and are drawn off to a subterranean channel at some point as yet undiscovered. This outlet has been sought for, and the logical place seems to be the southeast corner of the lake, where the basin is prolonged by the great valley of Bahr-el-Ghazal.

This is a dry valley, of uncertain slope and obstructed by dunes. For a long time the only portion of it at all known was the southwestern extremity where it connects with Lake Chad. Was this the outlet of the lake which, though dried up on the surface, continued to be active underground, in the depths of the soil? For half a century this was one of the problems of Saharan exploration; it was not solved until the Tilho expedition discovered an enormous basin some 435 miles northeast of Chad, and definitely determined the relationship between them.

The basin lies about 325 feet lower in level than the lake, and is hemmed in to the north and east by the high elevations of Tibesti and Ennedi. To the south and west however, no barrier separates it from the lake, and it is evident that extremely extensive lakes and morasses existed here at a fairly recent epoch. This is attested not only by the surface relief and the aspect of the terrain, but also by a very abundant subfossil fauna of mollusks and fishes identical with the existing mollusks and fishes of the present-day Chad. Tilho calls this basin the Chad Netherlands. The alluvial region of the Shari then, like that of the Niger, has receded by more than 400 miles, and for the same reasons: simply because such a region is essentially unstable; the river chokes up its own course by the accumulation of its own alluvium. But the waters of the Shari, like those of the Niger, continue to utilize the old buried channel, and thus contribute to the permanent aqueous reserves of the desert.

As in the case of the Niger, we must also consider here the part played by captures. Explorers, guided by native information, have sought for a communication by water between the Shari and the Benue, which would be of great practical importance. They have found that such a communication does indeed at times exist between the Logone, a tributary of the Shari, and one of the tributaries of the Benue, being established temporarily in flood periods, in a marshy region where the line of division is uncertain. [See figure, p. 71.]

This is assuredly a capture in the making; and the Benue, which belongs to the basin of the Niger, will eventually in the more or less distant future draw off from the Shari a part of its system. For the outcome of such a contest is never doubtful in the long run, when a lively opposition is set up between a vigorous fluvial system which has the sea for its base level, and a watercourse like the Shari whose erosive power is paralyzed by its alluvial region. When we have more knowledge of this locality we shall doubtless find evidence of ancient captures which similarly took place in the past to the detriment of the Shari, and which will furnish us with further explanations of its recession.

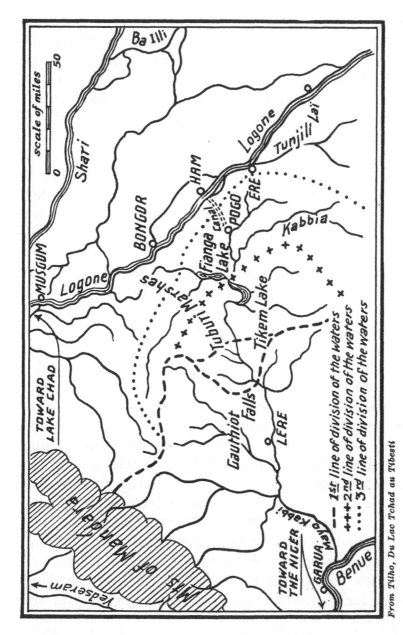

CAPTURE OF THE LOGONE BY THE BENUE

THE NILE

The Nile is in absolute contrast with the Shari and the Niger, for it alone succeds in traversing the entire width of the Sahara, and it thus carries the equatorial rains from Victoria Nyanza all the way to the Mediterranean. It is interesting to analyze the conditions which have made possible this triumph over the desert.

As everyone knows, there are two Niles: the White Nile and the Blue Nile, each of which has a definite and individual character of its own. The terms "White" and "Blue" are of accepted usage; but they are rather inexact translations of the two original Arabic words, and it would be more sensible, if not so literal, to say the Clear Nile and the Muddy Nile. These appellations would have a more direct bearing on the characters of the two rivers which have made such a profound impression on the popular imagination.

The White Nile is the one which would be an exact equivalent of the Shari and the Niger. It is typically the tropical river, coming directly from the southern hemisphere, with its sources below the equator. It likewise on reaching the Sahara spreads out in an alluvial region. The marshes of the White Nile are found throughout the region around the confluence of the Bahr-el-Ghazal * with the Nile, covering an extensive area immediately south of Fashoda between 28° and 30° east longitude and from 6° to 10° north latitude. They are laboriously drained by a great number of halting, entangled, deltaic rivers weaving through a network of morasses. These marshes of the White Nile have been known for nearly two thousand years, since the time of

* This is not the Bahr-el-Ghazal mentioned in connection with the Shari and Lake Chad. There are two separate systems of this name, which signifies River of the Gazelles.

Nero, when they were discovered by two Roman centurions who penetrated as far as this on an exploratory mission. It was in these same swamps that the Marchand expedition nearly came to grief.

The similarity of this region with the swampy regions of Timbuktu and Chad seems obvious. In the great plains of the southern Sahara any tropical river invariably halts, uncertain of its way, as if repelled by the desert influences. This retreat is in no way mysterious, for it is related to the mechanism of the erosion of an alluvial region. We may suppose that the White Nile, like the Shari and the Niger, if left to itself would terminate in its own sedimentary region. Certainly it does not emerge by its own force, without other aid. It too has been captured, and by good fortune this capture has been made by a powerful Mediterranean stream, the Blue Nile.

This is the Abyssinian river, fed by springs in the great massif of tropical Abyssinia where the summits reach an altitude of well over 13,000 feet. These mountains are not only an admirable source of water, but by reason of the steepness of their slopes they give to the Blue Nile its great erosive power, which is attested by the mud it carries. It is the Abyssinian loam which gives the Egyptian Nile its famous blood-red color during the flood season. Although the beginning of the flood is proclaimed at Cairo by the appearance of the green waters, freighted with plant debris from the marshy alluvial region of the White Nile, these green waters last but a short time, two or three days. As these waters are considered unhealthy, every household keeps a sufficient store of good water on hand until the red flood sets in, which water is fit for drinking purposes.

Now the last contribution from the Abyssinian

plateaus is received by the Nile at Berber, where it is joined by the Atbara, little brother of the Blue Nile. From Berber to the Mediterranean the Nile Valley has a development of over 1,400 miles. One would not naturally suppose that the impetus given to the river even by the steep Abyssinian slopes would be sufficient to carry it across so great a distance.

In seeking an explanation for so remarkable an accomplishment, let us first consider the shape of the valley, which is very simple. It describes an almost straight line in its entire length, and runs parallel to the great fault trough of the Red Sea. But it is strikingly asymmetrical, for the eastern side of the valley is formed by the Arabian range, a long sharply crested ridge with an altitude of from 6,000 to 6,500 feet; while the left or western bank of the river merges into the great Libyan plateau, which entirely lacks any outstanding relief and is of an altitude consistently less than 1,600 feet.

It is obvious that the general structure of such a valley cannot be the result of erosion. The river did not cut its own valley; it has merely followed one already made. What it has utilized was a long furrow in the earth's crust, a trench related to the same orogenic movements as caused the subsidence of the trough of the Red Sea. It has sometimes been called a fjord by Egyptian geologists, and this "fjord" evidently belongs to the great system of fractures which begins in the southern hemisphere with the Rift Valley of the Great Equatorial Lakes and may be followed northward through the Red Sea as far as the Dead Sea and the Syrian Coulee. Since the Nile has its source in the Rift Valley of the Great Lakes, it is not alone the fjord that belongs to this system, but the entire Nile Valley. In other words, to produce this phenomenon of a river victoriously traversing the desert from side to side, it has taken no less than one of the greatest accidents that ever happened to the earth's crust.

This great system of fractures, subsidences and up-heavals, which extends from the Rift Valley of the Great African Lakes to the Syrian Coulee, is not an extremely ancient feature on the earth's surface, at least not by the geological measurement of time. It is marked by volcanoes, some of which have retained the freshness of their original appearance, and some of which are even still active. Moreover, the valley of the Nile has the characteristics of a young valley; it is impeded in Egypt by the celebrated cataracts, showing that the river has not as yet had time to put its bed in order. Perhaps a closer examination of the valley will give some general idea as to the period when the present order of things became established.

We find the Arabian range to be scored by dead wadi valleys which seem exactly analogous to the Quaternary wadis of the central Sahara. These wadis converge with the Nile on its right bank and, in the pluvial epoch when they existed, could not have failed to contribute greatly to the work of erosion all along the valley. In fact, the present river, mighty as it is, would never have had force enough to carve out the Nilotic erosion valley as it is today without the aid of these Quaternary wadis.

Now although these wadis converge with the Nile on its right bank, none of them seems to have been able to cross it. They have no counterparts on the left bank. It was thus the valley of the Nile which withheld the benefit of Quaternary erosions from the Libyan Desert; by its negative influence, by the barrier which it imposed to the penetration of streams bringing the mountain rains, this deep fault is responsible for the Libyan

Desert of today. Thus when we seek to understand the Nile, we are forced once more to take into consideration that relatively pluvial period which preceded our own. In fact, the Nile itself, to a certain extent, is a Quaternary river; but one which, in contrast to the others, has the unique distinction of having survived.

THE ATLAS WADIS

The tropical waters are not the only ones to bring life into the Sahara. We must not overlook the rivers of the northwest which have their source in the Atlas, particularly the Moroccan High Atlas where the summits range around 1,300 feet. These are rivers of an entirely different nature. In the first place they are much less powerful. Even the Moroccan High Atlas does not constitute a water tower comparable to the tropical massifs which are lashed by the equatorial rains. Nor is the southern declivity of the Atlas, which is the only one we are to consider here, entirely free from the influences of the desert. The rivers emerging from it are never anything but wadis, in any part of their course.

On the other hand, the northern Sahara south of the Atlas is not a plain of indefinite slopes like the Sudan. The wadis of the Atlas are carried down slopes which are fairly steep, and for this reason even today in their flood season they are able to penetrate far into the interior of the desert. But there they remain. They offer a much better opportunity than do the equatorial rivers for studying the life of a desert watercourse, its struggle against contrary influences, its agony and its death.

There are two great wadis in the central part of the Moroccan Sahara, the Dra'a and the Tafilalelt; but as the Moroccan Sahara is still so little known, we can say only that each of these feeds a lovely oasis on the fringe of the desert, and beyond this we have practically no information. For a wadi typical of those originating in the Atlas we shall have to take the Saura, whose network articulates the Algerian Sahara, scattered with French outposts for thirty or forty years and consequently much better known.

THE WADI SAURA

In order to facilitate this analysis, we shall extend the appellation of Saura to include the entire wadi system, although the only channel which now properly bears that name and which is regularly active today is the most westerly one. It is made up at the little oasis of Igli by the meeting of its two principal arteries: the Wadi Susfana, which rises in the Saharan Atlas, and the Wadi Gir, which has its source in the Moroccan High Atlas, a much more important water tower than the Saharan Atlas of Algeria because of its greater elevation. From these heights come the great floods which at least once and often several times a year sweep the dry beds of the Saura from the mountains all the way down into the alluvial region in the vicinity of Upper Tuat. Thanks to the Saura channel, the beneficial influence of the Atlas snows and rains are carried right into the heart of the desert over a route of three or four hundred miles. If it were not for the Nile, this would be a unique phenomenon in the Sahara. The Saura is, in fact, a kind of little Nile, though of course on a very humble scale; it is certainly the only one of the peripheral wadis which can be in any way compared to it in regard to the power of its floods to penetrate into the heart of the desert.

Now the whole upper portion of the old Saura network remains admirably well defined, and for the rea-

son of course that the life of the wadis themselves is not
entirely extinct. Such rains as do fall in the vicinity,
however moderate, find a surface relief molded by Quaternary erosion to facilitate their flow; and while the
channels are dry the greater part of the time, they are
nevertheless periodically swept by the terrific floods
from the heights of the Atlas. Although this happens
perhaps not more than once a year or even more rarely,
it does happen sufficiently often to keep them flushed
and maintain them as active river beds.

These mighty flood bores, loosed perhaps by some
storm too distant to be seen or heard, race down the
channels like a tidal wave, carrying everything before
them. They give no warning and arrive so suddenly that
it is dangerous to camp in a dry wadi bed at any time, for
their unexpected appearance has often been the cause
of the apparently paradoxical occurrence of travelers
being drowned in the Sahara. It is for this reason that
French officers taking a detachment of troops into the
Algerian Sahara are instructed never, under any pre-
text, to camp in a wadi bed, however dry it may seem
to be.

The channels are naturally much less well preserved
in the heart of the system, where the sedimentary deposits of the converging tributaries arrested the process
of erosion. The establishment of a desert climate left
these accumulations of light alluvium at the disposal of
the wind, which has winnowed them and transposed
them into dunes. Thus we find in this region the great
erg known to the Algerians as the Occidental or Gurara
Erg, though it might perhaps be better named the Saura
Erg, since it obviously represents the central decomposition of the ancient Saura system.

In spite of this mask of the erg, the general plan of

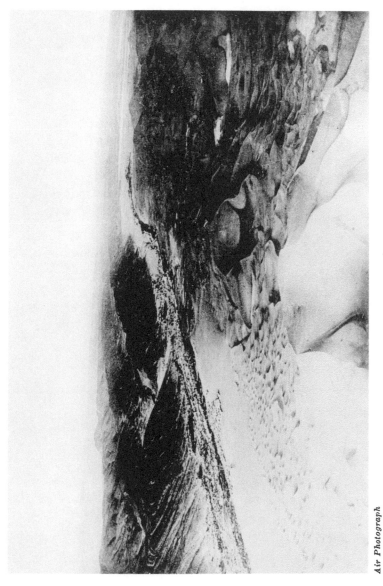

Air Photograph

THE WADI SAURA AT KARZAZ, SHOWING THE COURSE OF THE WADI BETWEEN
THE GREAT ERG AND THE DENUDED ROCKS OF THE UGARTA RANGE

the Saura stands out, distinct and clear as far as Tuat; though from this point on there is less certainty. Beyond Tuat the depression into which the Quaternary Saura must have flowed is occupied now by the Esh-Shesh Erg, a very arid and very difficult erg which is still almost entirely unknown and which guards its secret well. Up to this point, however, we can follow it, tracing with perfect ease the whole great tangled network of channels as they converge toward the bottom of the present basin, the lowest point of which is occupied today by the sabkha of Gurara.

It would seem that the mass of the intervening erg must close now to all the Atlas wadis any access to this, their former terminal; it is certainly deprived of any superficial running waters from the great floods. It is a very elongated, sinuous sabkha, bordered by high cliffs which show evidence of powerful erosion; and it is by following these cliffs that we can trace the principal bed of the old river as far as the oases of Upper Tuat. Timmimun, the capital of the province, from the top of the cliff overlooks the sabkha, which appears ordinarily as a barren lifeless plain, reddish brown in color. But there are some days when this plain is surprisingly seen to be covered with white spots which sparkle in the sun. This means that there has been a storm in the Atlas some time earlier; and while the flood, proceeding down one of the many channels of the wadis Namus, Rarbi or Seggeur, has been arrested by the dunes, the water itself seeping through the sand has finally reached the sabkha. Here capillary action has caused the salt to rise to the surface, thus demonstrating that even the buried channels of the ancient wadi system are not yet entirely dead. This whole process, as nearly as can be estimated, takes about a week.

It must be understood of course that the Saura itself, the active western branch, ends in an alluvial region; and this alluvial region is one that is well known.
The wadi decants into it after having made a sharp turn
to flow through the gorges of Fum el Kheneg, which are
carved in a ridge of hard sandstone at the southern extremity of the Ugarta range. And now, suddenly, the
flood no longer knows which way to go; for here the
alluvial delta begins. The wadi splits into two branches,
one continuing on directly south, the other turning
back and going toward the north. [See figure, p. 81.]

The southerly branch is not at all clearly defined; it
is only the heaviest floods which follow this course and
are able to reach the oases of Upper Tuat. Such an occurrence does not happen very frequently, but it has
been observed several times since the French occupation; and it is undoubtedly by this means that the
irrigation canals of these oases have been stocked with
the barbels, or fresh-water fish, that are found in them.

It is the most northerly branch of the delta which is
customarily followed by the usual moderate floods, and
this ends in a great closed depression which is known
as the sabkha of Timmudi, "terminal of the floods."
This sabkha has a very unusual aspect, quite different
from that of most sabkhas and shotts, such as the sabkha
of Gurara which we have just described. The salt in the
Timmudi sabkha is deposited in an almost pure state,
forming beds of rock salt comparable to the layer of
ice on the surface of a northern lake. The sharp variations of temperature cause this layer of salt, which is
several inches thick, to split into great irregular slabs
which pile up one on another like blocks of ice in the
polar ocean. The reason for this phenomenon is that the
water does not reach this place indirectly, by subter-

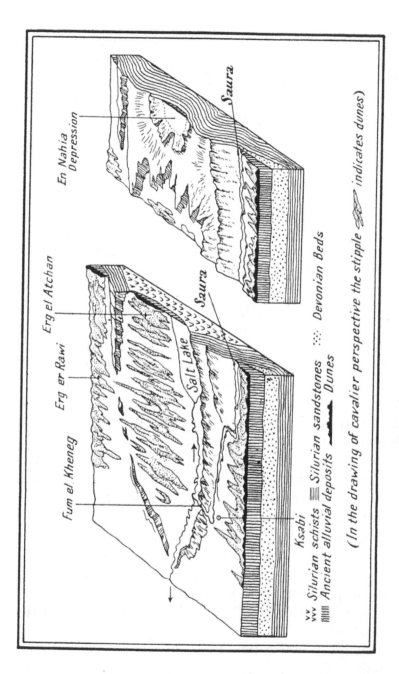

TERMINAL COURSE OF THE SAURA

En Nahia Depression

Saura

Erg el Atchan

Erg er Rawi

Fum el Kheneg

Saura

Salt Lake

Ksabi

vv Silurian schists ≣ Silurian sandstones ⠿ Devonian Beds
ⅢⅢ Ancient alluvial deposits ▲ Dunes

(In the drawing of cavalier perspective the stipple ⌇ indicates dunes)

ranean infiltration, as it does the other sabkhas. The flood arrives here in full force, and the whole volume of water spreads out on the surface and evaporates. Thus ends the Saura of today; and its ending explains for us the way in which many other great salt beds, like those of Taudeni, for instance, were formed in the past.

An exceptionally striking feature of the present Saura is the very curious asymmetry of its channel. The western side of the valley for a distance of something like 185 miles, practically its entire length, has a bulwark formation made up of lines of rocky hills. These are limestone above Igli and sandstone below; but both limestone and sandstone are bald, denuded rocks, swept clean and polished. On this side we have the rocky desert. The eastern or left bank, on the contrary, is followed by the edge of the great Gurara or Occidental Erg, so closely in fact that the Saura is really a trench forming the exact western boundary of the erg, which definitely halts on reaching it. [See figure, p. 83.]

The edge of the great erg comes into contact first with the Wadi Susfana at the little oasis of Taghit; and this contact once established continues practically without interruption as far as Fum el Kheneg. The fact is sufficiently curious to require a moment's attention. The channel of the Saura, although well defined, is nothing but a simple trench with no very great depth even at its deepest points; and it is perfectly dry for 340 days of the year. Can it be possible that such a secondary obstacle has been able all by itself to place a definite check on the progression of the great erg throughout the ages? On closer investigation it is evident that this is not the explanation.

At Taghit, which lies in the bed of the Susfana, at the point where the juncture between the erg and the

THE WADI SAURA AND ITS ERG

wadi takes place, an examination of the topographical conditions reveals the nature of the phenomenon which has been produced. Up to this point the Susfana, during its floods of course, flows in a Quaternary channel, an old valley bordered by terraces and having obviously a very ancient past. But at the little village of Zauïa Tahtania, at the lower end of the palm groves, the conditions abruptly change. Here we can see plainly where this old valley plunges beneath the erg in a direction almost due south, following what seems to be the general slope of the terrain in a course which would have taken the Quaternary wadi toward the point of convergence of the system, the sabkha of Gurara.

At Zauïa Tahtania the present wadi abandons its old channel, which the accumulation of the dunes has rendered impracticable, and works its way toward the west in the direction of Igli and the Gir, sliding as best it can between the edge of the erg and the calcareous cliff. In all this lower portion of its course the Susfana is a wadi without a valley, almost without a channel, its bed being simply a makeshift of the floods. Those that have force enough to push beyond this obstacle, which is not by any means the largest number of them, finally through their interrupted channel manage to reach the Gir, which is once more an old Quaternary valley, clearly defined. [See figure, p. 85.]

Although the relationship between the erg and the Saura has not been studied everywhere in detail, there are a number of points between Igli and Fum el Kheneg that show indications of conditions analogous to those observed between Zauïa Tahtania and Igli. In reality then, the present Saura does not flow in a Quaternary channel, but rather in fragments of ancient valleys, pieced together by accidental channels, so that the river

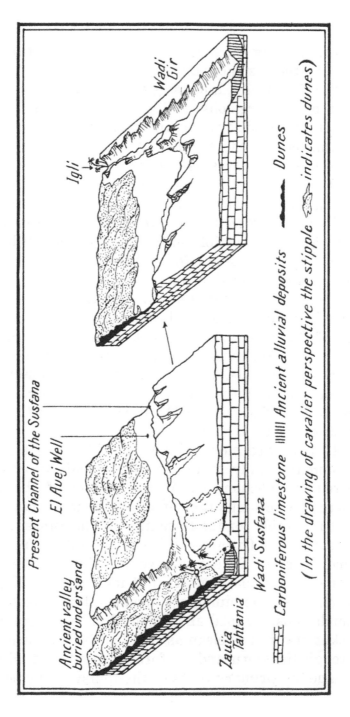

Present Channel of the Susfana

El Auej Well

Ancient valley
buried under sand

Zauia
Tahtania

Wadi Susfana

Igli

Wadi
Gir

⊞⊞ *Carboniferous limestone* ⦀⦀⦀ *Ancient alluvial deposits* ▬▬ *Dunes*

(*In the drawing of cavalier perspective the stipple* ⟋⟍ *indicates dunes*)

CAPTURE OF THE WADI SUSFANA BY THE GIR

system of today is the result of a series of captures imposed by the obstruction of the dunes.

The fact which impresses us here is the westward drive of the erg, obstructing and disorganizing the Quaternary wadi system, and pushing back along its external border the channel of the floods. If then the present Saura serves as the western boundary of the great erg, it is not because it acts as a trench to check the erg, but indeed for exactly the opposite reason. It has itself been forced back to the position where we now find it by the irresistible onslaught of the erg; it bounds the erg because the erg forces the floods to make a detour around its outside rim.

We may here note that this westerly drive of the erg is in evident accord with the present general direction of the winds. In this whole sector of the Sahara, where meteorological stations are plentiful enough to establish the facts, we know that the dominant wind is from the northeast or east-northeast, and related to the etesian winds of the Mediterranean and the trade winds of the subtropical zone.

If we now consider the opposite side of the erg, even a summary study of the terrain reveals inverse and correlative phenomena. At the bottom of the basin, where the accumulation of alluvium must have been most favorable for the formation of the dunes, we find the sabkha of Gurara—which is absolutely free from dunes. It seems indeed to have lost, through the action of the wind, a major portion of its original filling, for the cliffs of Timmimun are denuded, laid bare to their foundations, scraped and polished. The bottom of the sabkha itself is as clean as if it had been scoured, and through the worn and perforated mantle of alluvium the ancient primary rocks of the foundation emerge in great bald slabs. On this side it would seem that the erg had been in retreat throughout the ages, in the same measure as it had advanced on the other side.

Within the life and memory of man the erg is certainly immutable, but if we consider it in the light of geological measurement of time, this is not so. Then we see the erg in all its bulk moving bodily before the onslaught of the dominant winds. It is evident that the Gurara erg has noticeably shifted even since the end of the Quaternary age, which is to say, during the very geological period in which we now live. It tends even today, as a whole, to ascend the slopes of its basin toward the west, pushing back the Saura as it goes. [See illustration, p. 78.]

THE SAHARAN WADIS

It must not be overlooked that the Sahara has certain wadi sources of its own in the great mountainous massifs which rise to considerable heights in the very heart of the desert itself. These lofty ranges naturally also gave rise to some of the important Quaternary wadis, a few of which, in a very much reduced state, survive to the present day. For while these mountains are definitely in the arid belt, and consequently cannot be compared to the Atlas as a source of water, they do nevertheless receive more frequent rains than the surrounding desert, and thus maintain a certain degree of activity in some of the old channels.

The highest of the Saharan ranges would seem to be the Tibesti, for the altitude of Emi Kusi is estimated by the Tilho expedition as between 11,000 and 11,500 feet, which is some 1,600 feet higher than Ilaman, the crowning peak of the Ahaggar. The Tibesti is shown by the Tilho map to be scored with sharp valleys diverging

from it in all directions; and we know these valleys to
have perpetual water holes, since Tilho reported finding
a crocodile in one of them. But the ultimate destinies
of the wadis themselves, in the unexplored basins which
surround the massifs on all sides, have not yet been
determined.

The best known of the Saharan ranges is the Aïr, for
it has been seen many times since first viewed by Barth,
and there are some very good maps of it. It is much less
imposing in elevation than the others, but is nevertheless an important hydrological center for the divergence
of wadi valleys. Unfortunately, however, the only wadis
of the Aïr which are at all well known, even in general
outline, are the Sudanese streams, those whose valleys
trend toward the Niger. The truly Saharan wadis,
rising on the eastern side of the Aïr, are still entirely
unknown.

We are, however, sufficiently well informed as to the
Ahaggar and the wadis which descend from it; but here
too we must make a distinction. The Tafassasset and the
Tamanrasset are both Sudanese wadis. The Tafassasset,
once a powerful system but now almost completely desiccated, was first thought to have formed the original
headwaters of the Lower Niger, although it has just recently been determined that this wadi actually flowed in
the direction of Bilma, whose salt mines undoubtedly
indicate the location of the old terminal basin. The wadi
which joins the Niger below In Azaua has now been
found to be another, totally independent of the Tafassasset. The Tamanrasset in the Quaternary epoch seems
to have joined the Upper Niger in the basin of Taudeni.
In any case, we have no more than a rather general
and very imperfect idea of either of these rivers of the
Sahara.

THE OASIS OF IN SALAH : ARRANGEMENT TO INSURE EQUITABLE
DISTRIBUTION OF WATER AMONG ITS USERS

THE IGHARGHAR

We have, on the other hand, the Wadi Igharghar, which is not only very clearly defined but lies entirely in the Algerian Sahara. The whole network, but especially the northern portion which is the alluvial region, is in the midst of a section that has already emerged from the exploratory period and entered that of topographical surveys. It is therefore possible for us to analyze the Wadi Igharghar as we have analyzed the Wadi Saura.

In contrast to the Saura, whose original alluvial region is still unknown, the Quaternary wadi which was the ancestor of the Igharghar may be integrally reconstructed throughout, from the source to the delta. Its length was considerable, being something between that of the Danube and the Rhine. With its source below the tropic line and its terminal basin near Biskra, it had a development of some 620 miles in an almost straight‧ line. Its general slope was steep, since its head was at an altitude of nearly 6,600 feet and the bottom of its basin lay below sea level. It had an intricate and well-developed network of affluents, the outlines of which can still be easily deciphered between the frontiers of Tripolitania and the central ridge of Tademaït. It is probably the finest existing fossil of a Quaternary Saharan wadi known at the present time.

A glance at the general plan of the Igharghar reveals at the same time both the erstwhile cohesion of the system and its present dissolution. It is not difficult to reconstruct the Quaternary wadi in our minds, but such a mental reconstruction is nevertheless required. For parts of it have been destroyed and various features obliterated. The factor which has chiefly contributed to its present state has been the lack of a really dependable supply of water.

For we must remember that the Igharghar is the exact reverse of the Saura, in that it flows from south to north, rising in the heart of the desert and traveling toward its periphery. Instead of coming from the Atlas, it goes toward it. Now while the Ahaggar, by reason of its bulk and height, does attract some storms and has greater precipitation than the lowlands, it is nevertheless a desert range and is not now by any means an important water tower. Consequently the Igharghar has nothing at the present time which can compare with the consistent and mighty floods of the Saura that concentrate in a single channel and sweep it like a tide from end to end for 300 miles or more. Even if there were such floods, the very heaviest of them, coming from the Ahaggar and flowing through a channel whose continuity is extremely broken, could never possibly hope to reach all the way to the Great Shotts at the foot of the Atlas. Such a thing is not only improbable, but unimaginable.

Even in the upper reaches, in the Ahaggar itself, although we lack any systematic series of observations, there would seem to be no longer any common life among the branches of the network feeding the main artery. Each tributary of the system seems to have a fairly active life of its own; and while there are no real streams in the Ahaggar, there are certainly perpetual water holes, since we find them stocked not only with fish but with quite large ones. Apparently the most important of these pools are to be found in such wadis as have cut through the sandstone plateaus, which is natural since the standstone provides the best conditions for the accumulation and protection of subterranean reserves of water. In any case, the Mihero valley, where the Ahaggar crocodile was found, lies in a sandstone plateau; and it was in analogous pools of the Tibesti that the same species of crocodile was reported by the Tilho expedition. This identity of the residual fauna emphasizes the similarity of the general conditions in the Tibesti and in the Ahaggar, not only in the present but also in the past. It helps to establish the conviction that the Igharghar is representative of a whole category of Saharan wadis.

It is the alluvial region of the Igharghar which is particularly interesting. This lies at the foot of the Auras, the mightiest and best-watered range of the Saharan Atlas. All the rains that fall in the eastern half of these mountains, from Laghwat on, are brought into the basin by the Wadi Jedi, which circles the foot of the massif. Thus beneath the accretions of alluvium in the basin are imprisoned large reserves of water, which gush forth from artesian wells. Here in the oases of the Wadis R'ir and Jerid grow the finest dates of all the Maghrib. Altogether it is a corner of very great human importance. It touches on Algeria and is served by a railroad; there are good topographic maps of the country showing plainly its surface relief. This is one instance where the alluvial region of a wadi is better known than its upper course.

The heart of the system is naturally the part that has suffered most. The points of confluence, which were of course individual regions of sedimentation, are the points which show the greatest disintegration, and here again we find the development of the erg contributing to their detriment. For an accurate impression of the true character of the erg, we need only to glance at the map, which shows it for what it is: an all-consuming

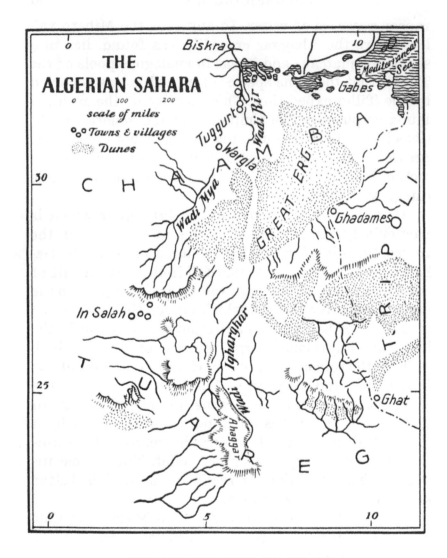

THE WADI IGHARGHAR AND ITS ERG

malady, a deformation, a sort of elephantiasis of the Quaternary wadi. [See figure, p. 92.]

The relations between the Oriental Erg and the Igharghar are exactly the same as those which have been determined between the Occidental Erg and the Saura; the similarity can be followed throughout. The erg of the Igharghar is decentered in relation to the alluvial region; while the bottom of the basin, around Tuggurt and Wargla, is practically free from dunes. Likewise the apparent displacement has been in the direction of the dominant wind, for the winter winds of this portion of the desert, under the influence of the Syrtes, are deflected and blow from the northwest, almost indeed from due north, while the erg appears to have been pushed forward on the east and southeast slopes, as far as the defiles of Ghadames.

VIII
Progressive Desiccation

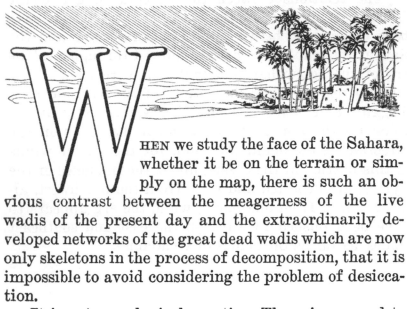

W HEN we study the face of the Sahara, whether it be on the terrain or simply on the map, there is such an obvious contrast between the meagerness of the live wadis of the present day and the extraordinarily developed networks of the great dead wadis which are now only skeletons in the process of decomposition, that it is impossible to avoid considering the problem of desiccation.

It is not a geological question. There is no need to ask whether the planet itself has undergone a tremendous change of climate in the direction of aridity since the Quaternary age. The fact is so self-evident as to allow no refutation. The question is entirely from the historical viewpoint: to determine whether during the short span of human memory the climate has continued to become less humid, and whether this process of desiccation is still going on at the present day. It is a question which has often been raised by geographers, not only relative to the Sahara, but to divers other regions of the globe. And up to the present time it is one that has not been definitely solved.

One of the regions which have been given particular attention in this regard is the interior of Central Asia.

Having been the point of origin of the great migrations which time after time convulsed the face of Europe, the theory has been advanced that the cause of these human crises may have been climatic. It is suggested that these overwhelming tides of yellow nomads, Huns, Mongols and Turks, which poured out of Asia and by repercussion possibly loosed the subsequent migrations of the Germanic tribes, may have been actuated by problems of desiccation which decreased the habitability of their homelands and drove them to seek more hospitable territory elsewhere. The question is still an open one, however.

As regards the Sahara, we are in a position to base our study on well-authenticated historical data, at least for certain sections of it. Along its Mediterranean front it has been associated with the most ancient memories of civilized humanity. Nowhere on the globe does history go so far back into the past as it does in Egypt, while the Maghrib has been in the full light of history since the days of Carthage, some two thousand years. Now the historians and geographers of antiquity describe the Sahara for us almost the same as we see it today; and while their descriptions, it is true, lack something in the way of scientific precision, nevertheless there are many actual monuments and relics of those early days from which we can gather data of an even greater exactitude. Nor have historians and archeologists in working with this material been able to bring forth a single positive fact which would permit us to conclude with certainty that there has been any change in the climate of these Mediterranean countries since history itself was born.

The evidence, on the contrary, points in the other direction, and is further borne out by an interesting agricultural experiment which has been taking place in the region of the Terres Sialines of Southern Tunisia since the French occupation. Some fifty or sixty years ago this was a steppe; but the ground was found to be littered with Roman millstones, testifying to an abundance of oil presses in ancient times. Putting faith in this archeological evidence, the Tunisian Department of Agriculture, under the direction of Paul Bourde, did not hesitate to undertake the exploitation of the springs and the planting of olive trees. In only a few years they have been able to produce crops which would seem apparently to duplicate the olive harvests of Roman times. Such a fact does not seem compatible with any real deterioration of climate since the Roman epoch. Furthermore, to quote Gsell:

Most of the springs which supplied the Roman settlements are still in existence. Has their volume diminished in the last fifteen centuries? . . . Such rare statements as we have, permit us to believe that in some places at least this volume has not been modified.

The negative conclusion then is the only one possible at the present time, in so far as concerns the actual climate, that is, the amount of precipitation. But if we consider the material desiccation of the Saharan terrain, we are faced with an entirely different question. It is obvious that the absolute quantity of the superficial waters within the bounds of the Sahara is constantly growing less.

The botanist Lavauden believes he has found evidences of quite recent desiccation in the Ahaggar region, at the very center of the Sahara; and the archeological excavations at the tomb of Tin Hinan in the same region would tend to suggest an aggravation of the desert conditions since the latter Middle Ages. There have also been discoveries as recent as 1934 of paintings and rock carvings in the vicinity of the Wadi Jerat, in the Ajjer Tuaregs' district of Tassili, which lead to similar conclusions; and these are definitely dated by their subjects, for they represent Garmantian chariots which cannot be more recent than the fifth century B.C. In this connection we might add that Passarge, in his study of the Kalahari, has collected impressive proofs not only of recent desiccation, but of the fact that the process is actually going on at the present time.

But we have already furnished the explanation of this fact, in showing how the Sudanese rivers like the Niger or the Shari, driven back by the increasing density of their own alluvial regions, or victims of capture to the profit of the ocean, have ceased to irrigate the southern Sahara. This deprivation may very well have taken place within historical times, or at any rate cannot be relegated indefinitely into the past. Moreover, the very process of the destruction of the Quaternary wadi systems also necessarily brings about a desiccation of the soil, as a summary analysis of this process will show.

CYCLES OF DESERT EROSION

Such rains as actually fall within the desert itself have a varying usefulness, depending on whether they find or do not find waiting to receive them an already existing network of valleys dug out by the erosion of rivers which have now disappeared. When no such network of wadis is there to organize the drainage, and the rain falls on rock of varying permeability, often slow or even nil, the enormous quantities of water loosed by a storm, only partially trickling off in short, disordered streams, are left to stagnate where evaporation is intense and almost instantaneous.

But a Quaternary valley, as long as it exists, forms a great natural irrigation canal. It not only serves in certain cases to carry into the heart of the desert the waters of distant rains which have fallen outside the desert domain, but in it the waters of the local storms are concentrated and carried swiftly to the alluvial basins, where they are imbibed by the lighter earth and form lasting reserves within its depths. The result is that all the vegetation is localized along the wadi beds or in their basins; in fact the words wadi and pasturage are interchangeable in the language of the nomads, who habitually reside in such places. This relationship between fluvial erosion and fertility holds good throughout, whether it be in the sandy or the stony desert. For the study of this phase of the erosion cycle, the best section is the Occidental Sahara, where the Quaternary networks are particularly well developed.

In the great erg regions the epidermis of the desert dune is perfectly denuded sand, forming a white or golden carpet which registers the passage of an animal or the fantasies of the wind in clear but ephemeral prints. Thus it remains, in spite of the fact that the sand is sufficiently permeable to absorb immediately and completely the waters of a storm. These quantities of stored water have no directly useful effect in the erg unless subterranean drainage concentrates them in favored spots, which then become pasturages. Thus such vegetation as there is, is determined by the Quaternary channels buried beneath the sand.

The two great ergs of the Algerian Sahara, the occidental and the oriental, both offer plentiful resources of wells and pasturages which they evidently owe to the buried networks of the Saura and the Igharghar. The occidental is the better known of the two, perhaps for

the reason that it is particularly furrowed by long lines
of verdure; and these fertile lanes are called wadis by
the natives, a term which is probably exact although the
valleys themselves remain indistinct beneath the dis-
guising contours of the dunes. The great Iguidi Erg of
the Moroccan Sahara is also in certain parts very hospitable, being strewn with wells and pasturages. The
water of course comes from some buried wadi, possibly
the Tafilalelt.

More curious is the contrast between the two small
distinct ergs found on the right bank of the Saura, one
of which is called El-Atchan, the erg of thirst, and the
other Er-Rawi, the humid erg. Of the two, only the latter has scattered wells and pasturages, for it follows a
valley which comes from a great distance, and the banks
of the wadi emerge here and there from beneath the con-
cealment of the dunes. The arid erg, on the other hand,
is inclosed in a wall of rocky ridges and is forced to rely
on its own resources of humidity.

In the stony desert the same connection is apparent.
In many places the wadi has no discernible banks, and
can be recognized only by the ribbon of green which
marks the course of the subterranean channel across the
surface of the plain. Or perhaps one may see only a
shallow bowl, a vaguely circular depression which alone
is carpeted with verdure in the midst of the surrounding waste. This is what the Saharan Arabs call a *daya,*
and is evidently the same as the "vleys" which Passarge describes in the Kalahari desert. We should here
note that the daya is not a terminal basin, for such
a formation presupposes indisputably, without pos-
sibility of contradiction, the presence of subterranean
circulation, since the water if it stayed in the basin
would assuredly deposit salt there, and we should have

a shott, a sabkha, or what is called a "salt pan" in
America and Australia. For water cannot remain fresh,
cannot be utilized by the plants, unless it circulates.

The great Lake Chad is of course an immense daya;
but it is not the only one which is well known. At the
foot of the Saharan Atlas, immediately south of Laghwat and almost in line with Algiers, there is a region
called the plateau of the dayas, where those verdant
basins alone give life to the absolute aridity of the surrounding desert. They are very beautiful dayas, irreg-
ularly strewn with splendid pistachio trees, which
although slender and growing at some distance from
one another, are altogether of a considerable number.
This very distinctive region used to be the favorite
habitation of the ostriches some three-quarters of a
century ago; but these of course, by reason of the European sporting frenzy, have now disappeared. [See illustration facing this page.]

It is not difficult to explain the daya plateau. It is
made up of the alluvial cones which have issued from
the Atlas over an immense duration of geological time
and have been collected into a single mighty mass. Consequently the soil to a great depth is extremely permeable. Although the whole plateau appears to the eye
to be entirely lacking in slope, it is really constituted of
a very sharp hogback or ridge between two large depressions which flank it to the east and west. Each of
these is a sink hole of subterranean drainage, an equivalent of what is called an "aven" [sink or swallet] in
France, or a "doline" or "polje" [depression in a limestone plateau] in the Balkan ranges. Beneath them
burrow the networks of the Saura and the Igharghar to
collect the water and organize its flow; in fact the very
form of the depressions has probably been determined

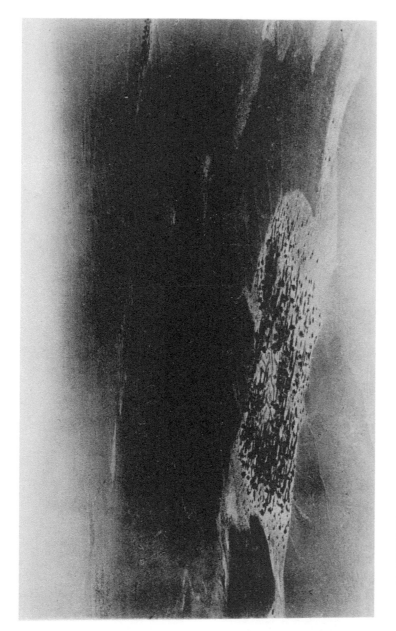

Air Photograph

THE "DAYA" PLATEAU, SOUTH OF LAGHWAT

by subterranean caverns which have caused the superficial crust to give way. Now these deep pits, although
so obstructed by silt as to be practically unnoticeable,
form a powerful attraction for the water that falls on
this plateau of light earth and indeterminate slope; and
this water, filtering slowly in the depths of the soil, has
determined the dayas for which the region is famed.
Thus their connection with the drainage organized by
the old Quaternary rivers, though less direct than in
other cases, is perfectly evident.

We have seen then how the river, long after its
death, remains beneficent by reason of the surface relief which its erosion has engraved on the face of the
desert. But this fluvially eroded relief is not everlasting.
It is ceaselessly attacked by the action of the wind and
there are constant effacement and decomposition. This
is shown plainly by a glance at the map, while a study of
the terrain itself permits us to formulate some measure
of the attacking force and to analyze the progressive
disintegration.

To evaluate it, let us first recall summarily how a
river constructs its valley. In the beginning it digs out
the hard rock; but farther on, in the lower reaches, it deposits its sediment in the basins, and its work is then
one of filling up. The result of this double, inverse process is the uniform slope of the thalweg; since, according
to the various sectors, it has alternately dug out and
filled in. After it is dead and the relief which it has created is attacked by eolian action, such portions of the
valley as are carved in the hard rock offer stubborn
resistance; but in the silted sectors the desiccated heaps
of light earth become immediately the prey of the wind.

The despoiling of the silted basins scatters all the
tenuous particles far and wide; everything which is of

a clayey texture is carried beyond the desert domain. The sand, sifted and winnowed, is transposed into dunes, swallowing up the channels, spreading over and roughly molding the irregularities of the terrain, and finally making up the great mass of the ergs. In the early stages of the desert relief, such as we find in the Algerian Sahara, the erg has a close relationship with the wadi systems and is dependent upon them not only for its own existence but for such life as it maintains. That this is in direct proportion would not always and everywhere seem to be true. For we have already cited cases, such as the Erg-er-Rawi and the Iguidi Erg, which offer plentiful resources of water and pasturage without apparent connection with any existing wadi; nevertheless the water must reach them through some buried channel, so that the wadi is not yet actually extinct.

But as the process of decomposition advances, the mass and extent of the erg are always increasing. Much of the substance of a given erg is carried away by the wind; but the migratory grains of sand usually go to enrich its neighboring erg, while eolian erosion is daily uncovering new beds of alluvium and transforming them into fresh dunes. Thus day by day the Quaternary channels are becoming more choked up, effaced and obliterated, while drainage becomes more and more difficult. Likewise the increasing thickness of the sand is also burying more deeply the underground supplies of water and making them more unavailable to vegetation and the uses of man.

This stage of the process is exemplified by the great Esh-Shesh Erg, by far the worst of the Algerian ergs and consequently also the least known. Its wells are rare, and one of them, Tni-Haïa, is poisoned with chlorine. It is one of the negative poles of life and seems to become more so day by day. The natives, within the memory of man, have ceased to bring their flocks here to graze, and they have even forgotten the trails; one can no longer find a guide for the Esh-Shesh Erg. That the eolian decomposition should be so far advanced does not seem unnatural when we remember that this erg covers what was once the lower part or terminal basin of the mighty Saura system, and that the fluvial erosion in this lower valley with its abating slopes was necessarily arrested from the very beginning by the earliest, Quaternary, sedimentation.

It would be unwise to attempt to formulate the details of this process too precisely. But the outcome of it we see in the Oriental Sahara. There, in the great Libyan Erg, we find what is probably the most imposing mass of dunes on the whole surface of the earth. They cover a vast area, some 750 miles in length by 250 to 300 miles wide. It is a region more unknown than the Antarctic, and unknown because it is impenetrable. Not only have European explorers been unable to enter its domain, but it is closed to the natives themselves. Considering that we have literally no knowledge of this Libyan Erg, the one thing that we can safely say of it is that by reason of this very fact there would seem to be nothing comparable to it in the Occidental Sahara.

Now, however, if we pursue our hypothesis to its logical conclusion, we must find that the wind tends to excavate these great alluvial basins again, and ultimately succeeds in emptying them entirely. In this way is created a strange landscape, disconcerting to the eye which no longer finds familiar lines to guide it. Chaotic upthrusts of the rocky skeleton, stripped of their alluvial padding, become unintelligible; there is a confused entanglement of fragments and interrupted cliffs. This is the *shabka,* a very expressive word signifying "network" or "fiber," which is used by the Arabs of the Occidental Sahara to describe this type of landscape.

If we now imagine this action prolonged not only for centuries, but for geological ages, we shall find immense areas where the sedimentation will have disappeared grain by grain, leaving perhaps only a residue of confused heaps of stones, so changed by corrasion that the original forms, the old fluvial markings, are almost impossible to recognize. Even the rocky skeleton itself, attacked by the multiple forms of desert erosion, worn down and blunted, will have taken on new outlines. Next suppose certain movements of the earth's crust to have intervened, as could hardly have failed to happen in so long an interval of time. They will have warped the pattern of the surface, and their effects will not have been counterbalanced by consistent fluvial erosion, as they would have been in normal climates. The result will be a surface relief like that of the Libyan Desert.

Here, in any place where the surface is not covered by the crumbling gravel of the serir, we find a plateau of rock, denuded and swept bare by the wind, gently undulating with shallow depressions in the form of vague hollows, yet deformed by sudden excrescences with abrupt slopes. In certain places, as for example in the vicinity of the oases, there are cliffs of a capricious design which seemingly cannot be explained in any wholly satisfactory fashion—neither by orogenic action nor by erosion, either fluvial or eolian. The true explanation would perhaps involve all three forces in proportions at present impossible to determine. For it has all been the outcome of a prodigiously long and arid past, during which the action of the wind has clearly taken the upper hand and has rendered the effects of the fluvial erosion indiscernible. In any case, this is the typical desert peneplane, absolutely desolate and deprived of all natural vegetation. The Egyptian Libyan Desert shows us the end of a labor which we saw in its beginnings in the Occidental, particularly the French Sahara.

When the idea of cycles was introduced, the study of the surface relief of fluvial erosion for normally drained countries was completely transformed. By means of this theory we have been enabled to distinguish the various forms as young valleys, those newly come to maturity, and those which are aging. The study of desert relief would seem to require an analogous base that would permit us to classify the two types which we find in the Sahara, making a distinction between the Occidental Sahara as a young desert, and the Libyan Desert as admirably representing the senescent type.

In any case, the facts themselves are patent. The marked difference between the two halves of the Sahara cannot be too strongly emphasized. The perhaps essential characteristic of this individuality is the extreme aridity of the Libyan Desert, incomparably greater than that of the Occidental Sahara. Does this indicate that between the two there is an actual difference in the climate itself, less rainfall and a greater degree of atmospheric dryness in the east than in the west? We certainly have no observations to that effect. But on the other hand, there are differences of surface relief that leap to the eye.

In the greater part of the Occidental Sahara the

Quaternary wadi systems are still in a remarkable state of preservation, while the Libyan Desert has retained no trace of these old fluvial erosions, although similar networks must formerly have existed. The presence of the serir alone would prove that they had, since its alluvial origin is obvious. But in addition to this, Egyptian geologists have recently discovered several enormous river deltas at Moghara, on the Wadi Natrun. These contain superb Pliocene and Miocene fossils, indicating that the fluvial period to which they belong is considerably more remote than the one which we have been studying. Likewise the valleys themselves have been entirely effaced by the wind and have completely disappeared.

At the same time, we find an extreme rarity or even in some cases a total absence of wells, pasturages and running water within the confines of the Libyan Desert. The only plausible conclusion seems to be that the terrain itself is materially more desiccated because the wearing away of the ancient surface relief of fluvial erosion is more advanced. This bears witness that a desert does become progressively arid by the simple material effect of the desert influences, without any change for the worse in the climate itself. It also, as we shall see, is immensely significant from the viewpoint of human habitation.

BIBLIOGRAPHY

CHUDEAU, R., "Sahara Soudanais," *Mission au Sahara,* Vol. II, Paris, 1909.

—— and HUBERT, Articles in *Annales de Géographie,* Vols. XXI, XXV, XXVII.

GAUTIER, E.-F., "Nomad and Sedentary Folks of Northern Africa," *Geographical Review,* Jan., 1921, p. 3.

GAUTIER, E.-F., "The Ahaggar: Heart of the Sahara," *Geographical Review*, July, 1926, p. 378.

—— "The Tomb of Tin Hinan," *Geographical Review*, July, 1934, p. 439.

—— "The Ancestors of the Tuaregs," *Geographical Review*, Jan., 1935, p. 12.

LAVAUDEN, L., "Sur la présence d'un cyprès dans les montagnes du Tassili," *Comptes Rendus de l'Académie des Sciences*, Feb. 22, 1926.

PELLEGRIN, J., "Les vertébrés aquatiques du Sahara," *Comptes Rendus de l'Académie des Sciences*, 1911, p. 972; *A.F.A.S.*, Tunis, 1913, p. 346.

TILHO, JEAN, Lt. Col., *Documents scientifiques de la mission*, Paris, 1910.

IX
Oases and Tanezroufts

DEEP-SEATED WATERS: THE OASES

W E have thus far attempted to analyze only the part played by the super-ficial waters of the Sahara, those which are located on or just under the surface, flowing in more or less organized channels and directly available. We have still to consider the artesian or deep-seated waters which lie in great beds beneath the rocky layers of the foundation and are brought to the surface only through the play of a fault or some rift in the crust. These are quite a different matter, but one of tremendous impor-tance, since in most cases they are the ones which are responsible for the oases.

There are, it is true, oases that are fed by rivers or by surface waters; in fact, the most magnificent oasis of the Sahara, and perhaps of the whole world, is one of these. This is Egypt, "The Gift of the Nile." It would seem to be a startling exception to the rule, al-though it is not by any means unique. But a good many of the Saharan oases, probably the majority of them, owe their origin to the deeper waters, often thermal, which are brought to the surface through faults or lines of contact between two geological formations.

The oases themselves are small, and few in number;

their combined area would give an extremely insignificant figure in comparison to the total expanse of the arid surfaces of the desert. Their great importance lies in the fact that they are places where the water, springing unexpectedly from the depths of the earth, flows with a paradoxical abundance. It must be, then, that there are important and permanent reserves within the depths of this dead land.

Where do they come from and how are they maintained? To arrive at some understanding of this, we must first try to imagine the inconceivable immensity of the arid wastes in which the tiny fertile oases are set. For the moment let us consider the Sahara as a small universe, with the vast sterile stretches representing the interplanetary or interstellar spaces. The surfaces of these immense areas are for the most part covered by the sandstone or limestone plateaus, the lava fields, the regs and serirs of light earth, or the sands of the dunes. What is probably by far the largest part of the whole extent is thus surfaced by these various types of terrain, all of which have some degree of permeability.

Next we must recall the fact that in any desert there is always some rain. The Saharan storms, rare as they are, each bring enormous quantities of water. One part of this is carried off by superficial circulation, while another is lost through evaporation. But a third and surely important part penetrates into the soil. Falling on these extensive and more or less permeable surfaces, some portion of the pluvial waters is slowly but constantly filtering into the depths, where at last it joins and maintains the permanent reserves.

This explanation alone has not been deemed sufficient by Passarge. He believes that these deep-seated reservoirs, like the networks of the dead wadis, date from the last humid period, the preceding geological age. According to this theory, the water that swells the dates in the palm groves would be Quaternary water, fossil water, contemporaneous with the prehistoric "Zambezi" fauna which has entirely disappeared. It is an amusing, grandiloquent idea; but one which has no value except for its suggestiveness, as is the case with any hypothesis in any science. It is at present impossible of proof, and is likely to remain so for a long time. But it has at least the merit of emphasizing the importance of the connection that links the present with the past, without which nothing in the sciences of nature is explicable or intelligible.

Now whereas the surface waters hitherto considered are particularly related to the watering places and pasturages, and consequently to the nomadic life, each emergence of these artesian waters forms a focal point of settlement and cultivation. The principal groups of oases are each individually distinctive and must be examined separately, as will be done later in detail. But as a class they represent the most extreme contrast with the desert conditions of the surrounding wastes; in fact, the very word "oasis" represents to the human mind something of paradise. Shaded by the branching palm groves, they nourish dream gardens that seem all the more miraculous after the long days of tedious and painful journeying across the endless dead immensities over which they are reached.

MAXIMUM ARIDITY: THE TANEZROUFTS

There is, however, a very great inequality in the distribution of the watering places, pasturages and oases throughout the Sahara; immense stretches are totally lacking in them and have no water resources whatever. Such regions constitute deserts within the desert—maximum or absolute deserts. These provinces of death are much dreaded by the natives, who are acutely aware of their menacing and dangerous character, and never fail to give each one of them some special name. But there is no term in existence to designate them as a class, and such a term is greatly needed.

Perhaps for our purpose we may be permitted to adopt and generalize a word from the Tuareg vocabulary, the name "Tanezrouft" by which they know a certain great portion of the Algerian Sahara which is completely lifeless. We shall here extend the term to include all such regions. It must be understood, however, that the tanezroufts are not all alike. The Tuareg Tanezrouft itself, in fact, is not of uniform character throughout. It extends from the summits of the Ahaggar to the Sudan, and the eastern portion, which has the greatest elevation, rises in a gently inclined plain from 2,000 to 3,300 feet. Here the dominant aspect is that of an old peneplane with worn, blunt ridges of more or less crystalline rocks piercing the thin mantle of the alluvial regs. Descending westward, this peneplane gradually disappears beneath the reg; until, in the vicinity of Taudeni on the meridian of Timbuktu, there is nothing to be seen but the endless level plain, strewn with gravel and without a single tuft of grass. There is not an undulation, nor any trace of erosion. It is all of an implacable uniformity, and the circle of the horizon is as regular as that of the ocean.

This is perhaps the most awesome and oppressive form of the tanezrouft. It is one which is also very prevalent in the Libyan Desert, particularly in the portion lying east of the Great Erg. The reg, or rather the serir, which extends inexorably across the whole distance between Cyrenaica and the oasis of Kufara is of this uninviting character throughout.

The Tuareg Tanezrouft is extended eastward toward Tripolitania by the Tiniri, which is its equivalent in sterility although it does not resemble it in formation. For this is what the Arabs call a hammada, a plateau of horizontal rocky strata, here made up of the characteristic Saharan sandstone, reddish in color with dark, almost black patina, which forms so large a part of the structure of the desert. The Tiniri sandstones are very old, belonging to the Devonian and Silurian periods. There are similar sandstones in the neighboring regions of Tassili and Muidir, but these are hospitable to man, being strewn with watering places and marked by pasturages, for the reason that they have been squeezed, sunken, scored with long canyons, and hollowed out into basins. [See illustrations, pp. 112, 164.] But in the Tiniri they are uniform and implacably horizontal, being almost as flat as the reg and as lacking in valleys.

At the opposite extremity of the Sahara, in the southeastern part of the Libyan Desert, on the left bank of the Upper Nile, there is also a kind of "Tiniri" or similar formation. Here the Nubian sandstones, although geologically very much younger, being only of the Cretaceous age, present almost the same features as the Silurian and Devonian sandstones of the Tiniri itself, and offer the same uniformity of horizons.

Drawing toward the Mediterranean, the Libyan Desert becomes a calcareous or limestone plateau. In places its surface, or its epidermis rather, bristles with a fine and compact lacework of sharp ridges pitted by corrasion, a formation analogous to our lapies and pro-

Photograph by Désiré

THE GORGES OF ARAK AT MUIDIR

duced by the action of the infrequent rains combined with that of the wind. It is called *kharafish* by the Egyptians, and is exactly the same thing as Sven Hedin describes under the term *yardang* in the deserts of inner Asia. To travel across such a terrain is torture, almost an impossibility. But aside from details of this nature which differentiate the various sections, the Libyan plateau, whether of calcareous or sandstone formation, retains its same unbroken horizons and its uniform character of tanezrouft throughout.

In other places it is the ergs, the great dune regions, which are tanezroufts, inherently lacking in hospitable resources. The great Libyan Erg we have already shown to be of this nature; as also the Esh-Shesh Erg in the Occidental Sahara between Taudeni and the Saura, which, although accessible at a pinch, is far from habitable. To these we must also add the awesome Juf in a neighboring region northwest of Timbuktu, of which little is known except that it is an immense shallow depression, as is indicated by its name, which signifies "The Belly"; and that the erg forms a large part of it. Its eastern extremity of course is fairly well known, thanks to the exploitation of the salt works of Taudeni and Teghazza; but what is by far the most considerable portion of it extends westward from here, beyond any of the caravan trails.

These various forms of tanezroufts, although differing in feature as they do, may well have one point in common. They are perhaps much older deserts, and more arid than the other portions because they have had more time to dry up. Certainly all of them show a characteristically senescent surface relief. We must, however, be cautious about forming hasty generalizations and mathematical deductions, leaving always a wide margin for the infinite complexity of the phenomena in the sciences of nature. What we must try to do is to get a clear and comprehensive idea of the importance of these tanezroufts, not only as regards size, but as a powerful obstacle to life. In truth, all life is banished from them absolutely; within their confines no one lingers, merely passing through. Except for the trails that cross them, they have no interest for the inhabitants of the desert.

In the Occidental Sahara they are rather a hindrance than an obstacle, for the main caravan routes cross them in many places. The most frequented route between Algeria and the elbow of the Niger crosses the Tanezrouft between the wells of In Ziza and Timissao, where we have a stretch of 110 miles without water or pasturage; and coming from the south there is almost the same distance between Timissao and Silet, the first watering place on the borderland of the Ahaggar Mountains. To our occidental conceptions, a distance of over 100 miles without water seems appalling; but to the Saharans, making all haste from point to point, it seems no such great matter.

Farther to the west, however, the Tanezrouft widens out and acquires its full power, so that between Lower Tuat and Timbuktu any direct communication is almost impossible. After leaving the Wallen well, the last stopping place within the Algerian domain, there is a distance of 325 miles entirely unprovided with permanent watering places before reaching Ashurat, the first halt in the Sudan. This in ordinary times is a closed route, although in rainy years it becomes accessible because it then has certain temporary pools which are well known and carefully landmarked.

But it is in the Oriental Sahara, in the Libyan

Desert, that the tanezroufts become an almost insurmountable obstacle. On the caravan route between Buttafel and the first watering places of Kufara there is a stretch of something like 185 miles without a single blade of grass or a single drop of water at any time. The caravans which regularly frequent this route generally allow seven days to cross it. Nowhere do they see anything but a uniform reg, in which they must find their directions by the stars. And the oases at either end are only of moderate size; they might easily be missed and passed by if an error of direction were made. It would certainly seem to be the most dreadful route in the Sahara, at least of those in frequent use. Rosita Forbes, who followed it from Cyrenaica to the oasis of Kufara, gives in her vivid narrative a very clear idea of the hazards of such travel. She notes by hearsay another route in this same region, connecting Kufara with the Egyptian oasis of Farafra on the opposite side of the Libyan Erg. It would seem to be an even worse one, and is but rarely used. No European has ever followed it, and according to the natives it requires a march of twelve days without water, across difficult dunes from one end to the other.

It is in these empty, dead immensities that the desert dangers lurk. Man's imagination seems to have a special predilection for picturing that of the simoom, and caravans are described as being thrown to earth by the wind, swallowed up and drowned by the moving waves of sand. But this is a purely literary conception, for no sand storm, however impressive and troublesome, has ever destroyed anybody.

There are, however, real dangers in the desert; and one of the lesser ones which should be pointed out is that of poisoning, which is unlooked-for. Occasionally the water of the wells is disagreeable to the taste, even nauseating, and often purgative; but in a few cases, which fortunately are very rare, it is also charged with noxious salts, especially chlorine, which may indeed be strong enough to cause death. According to Laperrine, the water of Tni-Haïa in the Esh-Shesh Erg is of this variety. It burns the clothing, and causes those who drink it to bloat. All the officers and soldiers of his detachment had their hands and feet more or less swollen, and in the case of one young native soldier the edema lasted for a month. Laperrine in the same erg also encountered water which had such a high content of saltpeter that it caused those who drank it to vomit blood.

This danger of poisoning, it is true, should be cited only as something in the nature of a curiosity. The true danger of the desert is death by thirst. Even this in reality is not so terrible as one would imagine it to be. In the agony of thirst, consciousness appears to be lost long before death. On this subject we have the testimony of that excellent observer Barth, who was dying of thirst in the Tripolitan Sahara when he was found by his companions and revived. He describes his principal sensations as being a semiconscious coma and an impotence to move.

Laperrine tells a story of certain native meharists who had had no water since the morning of the previous day. Through the false pride of the Saharan, and spurred on by legends of some famous robber who, like his camel, was wont to go for two and three days without water, they made no complaint. But in the afternoon the thirsty ones fainted, and had to be revived by subcutaneous injections of caffeine, and by being made to swallow tiny mouthfuls of water.

It is thus that death takes its toll in the Sahara. General Laperrine himself, after an airplane crash, came to his end in this manner. Nor is it rare to find the little-frequented Saharan trails strewn with victims, half-mummified by the dry desert air, waiting for a month or two the charity of burial. Rosita Forbes on the route to Kufara saw "a group of skeletons still fresh, evidently the remains of a whole caravan who had died of thirst." The bodies of those who stray from the trails are never found, and they are reported only as having disappeared.

This always present danger has taken a powerful hold on the human imagination. Consider the departure of a caravan setting out on a route where it is known that so many others have perished, and having no other directions than these: "Keep the pole star in line with your right eye, and march all day until you have sighted the evening star"; with the additional counsel: "Above all, do not bear too far to the west, or you will go to the devil." Imagine the interminable journey across the unbroken reg, day after day, hoping always for the mirage, because the mirage enables one to see beyond the horizon and may perhaps show some landmark at a greater distance to give the direction. Think what the sensation of the traveler must be when less than a pint of water remains for seventeen persons, and the guide has manifestly lost the trail, and the more hot-headed members of the caravan are eyeing him furtively and fingering the butt ends of their rifles. In such critical moments, the Saharan natives know the danger of emotion; and in one of their legends this danger is personified.

The desert also has its voices. The abrupt transitions from darkness to daylight are often accompanied by

the shattering of the desert rocks, with a grating sound
or a loud noise. According to the ancients, it was thus
that the Colossus of Memnon saluted the dawn when
touched by the first rays of the sun. The dune likewise
talks, for on certain days in certain dunes the stirrings
of the wind, or even just the pressure of the human foot
will cause shocks and tremblings; and then the myriads
of grains of sand, rubbing gently one against the other,
will make a strange snoring noise not unlike the rolling
of a drum. To the natives, these mysterious noises are
the bursts of laughter of a jinn, one whom they call Rul,
and who is the bad angel of strayed travelers. When the
wanderer has lost his way, when the exhaustion of
fatigue and the lassitude of thirst and the anguish of
danger are beginning to confuse his sight and paralyze
his brain, then he is tormented by the laughter of Rul.

BIBLIOGRAPHY

BEADNELL, H. J. L., *Dakhla Oasis*, Cairo, 1901.
—— *Baharia Oasis*, Cairo, 1903.
—— *Farafra Oasis*, Cairo, 1901.
FOURTAU, R., *Vertébrés miocènes de l'Egypte*, Cairo, 1902.
Geological Survey of Egypt, maps and monographs.

PART III
HISTORY OF THE SAHARA

X

Introduction of the Camel and Its Consequences

I N one respect the Sahara is the antithesis of the North American and Australian deserts, which were found by European immigration in a virgin state. For the Sahara, while certainly never a densely populated region, has always in a certain measure been the home of man. Thanks to the works of Reygasse and other prehistorians, as mentioned by Henri Breuil, we now have the certainty that it was inhabited as early as the Quaternary age; and that man has been present during all the long process of desiccation which has changed the Quaternary Sahara into the Sahara of today.

These early desert dwellers have left behind them stone weapons and utensils which can be very exactly identified by the accepted European classifications as Chellean, Acheulean, Mousterian and Solutrian. Moreover, as we might expect, the most archaic of these implements seem to be found in the more desolate regions; while the examples of finer and more highly developed workmanships are limited to such portions of the desert as still bear some relationship to present-day life—showing that as certain places became uninhabitable, they were abandoned for others where conditions were more favorable. Enormous rude hand axes, for instance, recalling our Paleolithic forms, have been found in great quantities in the Esh-Shesh Erg, where as we have shown, the desiccation of the terrain must have begun very early; while in sections where the desiccation is less advanced, the ground is littered with sharp spearheads and delicate Neolithic utensils.*

But all this is in the realm of prehistory, and therefore a little nebulous. It is naturally the historical era which interests us more; for the Sahara, linked with the ancient Mediterranean and intermingled with the histories of Egypt and Carthage, of Rome and the Arabian Empire, has a history of its own that takes form as we study it. The principal and illuminating fact in this desert history is the introduction of the camel, who put in a very tardy appearance. Today the camel, or to be more exact, the dromedary—the camel with one hump—is so closely associated in our minds with the desert landscape that the two seem inseparable; but in actual fact the camel is comparatively a newcomer. He is never mentioned by the ancient historians, nor does he figure in the rock engravings and pictorial monuments which record for us so exactly the everyday life of these early peoples. Nowhere does he appear in the picture records which show us the history of all the thousands of years of Egypt's independence; and he was certainly not in use in Punic and early Roman Africa during the times of Sallust and Pliny the Elder.

In the days of Carthage and the first part of the Roman Empire, it was the elephant that was used along the border of the desert in somewhat the same capacity as the camel is today. This seems almost paradoxical,

* A map showing the arrangement of the lithic beds in the Sahara was exhibited by Reygasse at the Saharan Exposition in the Trocadéro in 1934. This map has not been published.

for the elephant is by no means a desert animal; nevertheless it is a well-authenticated historical fact. We know that there were bands of wild elephants which roamed the Atlas Mountains and in winter came down into the humid Saharan basins at the foot of the range. History tells us of Hasdrubal, in order to replenish the elephant reserves of Carthage, setting out to capture wild elephants in the basins of the great Tunisian Shotts and the Wadi R'ir, where the palm groves of the present were not as yet in existence. Nor was it until well into the Roman period that the wild elephant disappeared from the Atlas, wiped out by the economic exigencies of the Roman market with its demand for ivory, as well as by the destructive fury common to the European of all times.

These elephants, as we have shown, were small of stature, presenting the characteristics of degeneration of the residual fauna to which they belonged, just as do the crocodiles of Tassili and Tibesti, or the fishes and cobras of the Wadi Igharghar today. Assuredly they had long been separated from Equatorial Africa, their country of origin; for the desert trails of the interior Sahara of course were closed to them.

It would seem that man at this time used pack oxen when he did journey on these trails, for the ox, with a kind of packsaddle on its back, figures often in the rock engravings. Nor is the Sahara even today entirely closed to the zebu ox of the Sudan, for the Tuaregs of the Ahaggar, forever shuttling back and forth between their mountains and the Niger, take some of these zebus with them in their continuous peregrinations. The animals are provided for by having them carry on their own backs the water and forage they will need in their wanderings. In some parts of the Sahara the horse was also used; at least we can be certain that it was not unknown in Phazania, or what is today Fezzan. For the ancients have left us records of raids in which horse-drawn war chariots figured, apparently of the same model as the Pharaonic chariots pictured on the Egyptian monuments. Moreover, the Gautier-Reygasse expedition of 1934 found ochre paintings and rock carvings in the Ajjer Tassili which show us that these war chariots, drawn by two and three horses, were in use even in the Tuareg mountains.

But nowhere in all this time is there any trace of the camel. It was first imported, according to Egyptologists, during the Persian conquest in 525 B.C.; and from that time it played a certain rôle in Egypt, particularly for communications between the Nile and the Red Sea. But its use was confined entirely to the east; it was not until centuries later that it penetrated into the occidental desert. This is not surprising when we consider to what an extent Egypt is set apart from the rest of the Sahara.

The introduction of the camel into the Occidental Sahara took place during the time of the Roman Empire, but particularly toward the latter end of it. It appeared first in Tripolitania, as was natural, since the Punic-Roman cities of that country, especially Leptis Magna or Tripoli, had always been supported by the trans-Saharan traffic which came through Ghadames, Phazania, and the city of Jerma-Garama, capital of the Garmantians, who were perhaps the ancestors of the Tuaregs and even at that time a caravan people. By the time of Ammianus Marcellinus camels were found in Tripolitania by thousands; and in Byzantine Africa at the period of Procopius and Corippus they were already established in their present status and playing

Photograph by Gautier

CAMELS EXPORTED FROM ALGERIA TO EGYPT DURING THE WAR

an important part as beasts of burden and companions in war, exactly as today. From this point the evolution of their present rôle was an accomplished fact.

A question which naturally comes to mind in this connection is the old one of progressive desiccation; for when we find the camel substituted for the elephant we can hardly refrain from suspecting that the substitution was influenced by a change of climate. This is an entirely natural supposition, since we know that the desert, independently of any diminution of the general rainfall, is automatically self-desiccating by the simple prolongation of the desert conditions. And in a country like the Sahara, already so close to the absolute limit below which any form of life would be impossible, even a very slight lessening of the average precipitation can bring about far-reaching consequences. It would therefore be unwise to discard a priori such a normal hypothesis as that of desiccation.

But it is, after all, only a hypothesis, and not in any way indispensable to our understanding of the observed phenomena. The last three or four centuries, which we may consider the modern historical period, have witnessed an amazing transformation on the face of the globe; almost under our very eyes the new worlds of America and Oceania have been inundated by waves of mass immigration of alien races and fauna. Yet it does not occur to us to seek for some climatic crisis as an explanation, for we know these to have been purely historical events. Why then should we not believe that in the more distant past such a tremendous historical reality as the Roman Empire should have been sufficient in itself to explain a transformation in the domestic animals of North Africa? Especially does it seem plausible when we consider a man like Septimius

Severus, who was born at Leptis Magna in Tripolitania, and imbued with the traditions of a race which had always lived by trans-Saharan commerce, remembering that at one time this African, this Saharan, held in his hand the entire military, political and economic power not only of the Roman Empire, but of the whole Mediterranean world.

We cannot say whether the introduction of the camel, coinciding so nearly with the close of the Roman Empire, is simply a chronological fact, or whether we should read into it some deeper significance of cause and effect. However this may be, the importance of the fact remains; the appearance of this animal proved to be the turning point in the history of the desert itself, the great event which caused a radical transformation in its destiny. Historically we may say there are two Saharas: the precameline Sahara, and the modern or cameline Sahara.

The White Nomadic Conquest of the Desert

Prior to the advent of the camel, the scanty population of the desert was of the black race. Egypt of course must be excepted; but aside from the Egyptians the only other whites were the Berbers, and they occupied only the fringes of the desert through Tripolitania, the Maghrib and the Atlas regions. According to the unanimous testimony of the ancient writers, the Sahara itself was the domain of the "Ethiopians." They extended from Libya to the Atlantic, and as far north as the Great Shotts.

According to the early historians, cited and commented upon by Gsell, the frontier between the Berbers and the Ethiopians lay at the foot of the Atlas, exactly along the line of the Wadi Jedi, which has its source in the vicinity of Laghwat and terminates in the basin of the Great Shotts directly south of Biskra. But on this point we have testimony even more unassailable than that of the texts, for throughout this region, including the Wadi R'ir and the dunes of the Lower Igharghar, the ground is found to be strewn in incredible profusion with the finest examples of Neolithic industry. In this remote and isolated corner, the use of such weapons and implements was certainly extended into the historical era; and even now, to be exact, has not been entirely abandoned.

What is most striking about these finds is the abundance of arrowheads of an admirable workmanship and finish. Now the bow and arrow have never been utilized by the Berbers, whose only casting weapon has always been the zagaya. Not only is the testimony of the ancient texts unanimous and irrefutable on this point, but it is also shown by the present practice of the Berber Tuaregs, who even today, around Timbuktu and the elbow of the Niger, throw the zagaya with atavistic skill while mounted on horseback and riding at full gallop. Also throughout the military history of the Maghribi Berber Sultanates, whenever there is any mention of a corps of archers it has always been a troop of Asiatic mercenaries, particularly Chorasmians. But the bow and arrow are and always have been the traditional armament of the Nigritians, and are used today by the Hausas in Aïr, where they particularly caught the attention of Foureau, being a novelty for one coming from the north.

There are other equally significant implements of polished stone which are found in great profusion in the Saharan soil. These are huge rollers and great bell-shaped mortars of a type not only well known to archeologists and prehistorians, but still in use in the Sudan. They served the early inhabitants to crush grain and reduce it to flour; and often they are found in places now far removed from any human habitation, but often also near the present palm groves, as in Tidekelt, where because of their elongated form they are utilized as funeral steles (*shahad*) in the Moslem cemeteries. But in the domestic life of the Sahara their place has now been taken by the small Mediterranean hand mill or grindstone. Besides, cereals have now been reduced to a subordinate place in the nutrition of the country, having been to a great extent supplanted by dates, with meat and dairy products furnishing the necessary variety among the nomads. Thus rollers and mortars evidently date back to a preceding period, that of the Saharan Negro. They give us a clear picture of the past, of a Sahara in which the white race, the camel caravans and the palm groves, all now so closely associated, had not as yet put in an appearance; they serve to show us not only the importance but the nature of the transformation which has been accomplished.

During the past fifteen hundred years, since the Hausas scattered their arrowheads about the terminal basin of the Igharghar, there has been a great thrust from the north in which the white Mediterranean races have never ceased to drive back the Negroes. This onslaught has been less violent and less efficacious in the Oriental Sahara than in the western portions of the desert, and in fact the penetration of the white race into the Sahara has on the whole been a leisurely and progressive process—one, indeed, which is still continuing, as we might say, practically under our eyes.

The scene of the earliest and consequently the most interesting phases of this progressive conquest of the desert was the Algerian Sahara. The whole group of oases which constitute the backbone of the Algerian Desert is relatively recent, including the beautiful ones of the Wadi R'ir, so close to what was Roman Africa and so prosperous today; as well as those of Gurara which lie a little farther from the Atlas, but are easily accessible. In spite of their proximity to the old Roman boundaries, they were never mentioned by ancient writers, and in them is found no archeological trace of Rome. They are obviously not of Roman origin; in fact we know them to have been founded toward the close of the empire by the Zenetic Berbers, who were more or less Hebraic.

These Jewish Zenetes were great nomadic cameleers who were thought to have had some connection with the famous Jewries of Cyrenaica that gave the Roman Empire so much trouble. They would seem to have arrived on the scene during the sixth century A.D., for native traditions recorded in Arabic chronicles have specially preserved the memory of a great immigration in the famous "elephant year" of the chronology immediately preceding the Islamic chronology. We have therefore a date which can be fixed with some certainty, or at least a sufficiently close approximation, for the foundation of the palm groves of Gurara and Upper Tuat, which is the gateway into the Algerian Sahara.

These Zenetes had their capital at Tamentit, and maintained themselves in Gurara and Upper Tuat until the sixteenth century, playing an important part in Africa Minor throughout the Middle Ages. They have left in these regions not only funeral pedestals inscribed with Hebrew characters, but also some living descendants; their name and dialect (Zenatiya) are still closely associated with Gurara even today. Yet, strangely, their name, unlike that of many less illustrious Berber tribes, is never mentioned by the ancient writers. It seems most significant that their appearance, precisely at the end of Roman and the beginning of Byzantine Africa, should also coincide exactly with the very moment when the camel first appeared in great numbers within the confines of Africa Minor. It is difficult to escape the idea that there is some link between the two events; the nomadic Zenetes, dependent upon the camel, would seem either to have brought it with them, or to have come simultaneously with its importation.

The thrust from the northeast was carried only very slowly into the interior of the Sahara. The foggaras, or oriental irrigation systems, of Lower Tuat, and consequently the palm groves as they exist today, go back only to the third century of the Hegira, which corresponds to our tenth century A.D.; while the political domination of this region was retained until about the fourteenth century by a Sudanese tribe related to the Bambara group, although more or less crossbred with the Berbers. Not only is this affirmed by native tradition, but ruined villages said to have been built by these early masters are still to be found throughout the region. They are quite different from the present villages in both architecture and arrangement; and they are also situated much farther from the present palm groves. Going still farther south, we find at Tidekelt that the very oldest groves do not date earlier than the thirteenth century A.D., while the more recent ones came into existence as late as the eighteenth century.

Certain other stages of this conquest can also be determined historically, although in some regions it is easier to trace them than in others. One of these is

MEHARISTS OF ONE OF THE SAHARAN COMPANIES IN UNIFORM

Fezzan, since this was the Phazania of the Greek and Latin authors and has not changed its name since earliest antiquity. It was the country of the Garmantians, whose name is reflected in that of Jerma, one of its oases. Duveyrier, making use of native traditions to supplement the evidence of ancient historians and Arabian chroniclers, believed he had established the identity of this people as a Negroid race related to the Bornuans; but the 1934 discoveries of ochre paintings and rock carvings in the Wadi Jerat of Tassili have forced us to revise our ideas. We are now led to conclude that the Garmantians, at least in the beginning, were Mediterranean colonists of the white race, and had become gradually negrofied over a long period. It would seem that this is the strain which has survived, partially at any rate, in the present Tuareg tribes. In any case, Fezzan, although partly under Cathaginian and Roman rule, remained until its conquest by the Bedouin Arabs an empire of the same type as the great Nigritian and Sudanese empires found by Barth on the banks of the Chad at a much later date.

There were similar great Nigritian empires throughout the western Sahara during the Middle Ages, and those of the Ghana, Sonroï and Manding tribes were important enough to attract the attention of the Arabian chroniclers. It is also now thought possible that the Negroid population of the Sahara included the Bushmen, since many of the ochre paintings found in Tassili, at Ghat, and In Ezzam, as well as at Aïn Dua in Upper Egypt, strikingly resemble those of the Bushmen of South Africa. The western Sahara, however, was the section which was chiefly overrun by the Berbers, and such incidents as the conquest of Timbuktu by a Moroccan army in 1591 emphasize how deeply they had already penetrated into the interior of the desert. Even the capitals of two of the most powerful Nigritian empires, Ghana and Gao, now in ruins, were located on the borderland between the Sudan and the Sahara, in the vicinity of Timbuktu. By the time of the French conquest the whole region around the bend of the Niger was found to be completely dominated by the Tuareg influence, not only from the political and economic points of view, but racially as well. The Berbers by now have eliminated not only the Nigritian influence, but indeed practically the entire Nigritian population.

Likewise in Aïr, which is really an intermediate step between the Sahara and the Sudan, the Tuareg conquest already partially overshadows the fundamentally Hausa population; while in the Libyan Desert at the eastern end we find another interesting and absolutely incontestible fact which proves once more that the conquest of the Sahara is still in the process of accomplishment. The word "Kufara" signifies "The Pagan," which would seem to be a curious name for an oasis that today, in the impenetrable fastnesses of the Libyan Desert, is the capital of Senusism and the last retreat of independent Islam. The origin of this name is historically known and dates back only a century and a half, for it commemorates the victory of the Moslems over the Tibbus, the Tibesti Negroes who had held Kufara up to that time as a fortified outpost.

The ruins of the Tibbu villages are still plainly to be seen, as is natural in a land where life is not intensive enough to obliterate the traces of the past; and in fact the last of the Tibbu aborigines may still be found there, though in a subservient position and in ever-decreasing numbers. This then is a particularly

clear case in proof of our point. Kufara was the last
stronghold of the black race in the Libyan Desert, and
it was conquered by the whites only toward the close of
the eighteenth century. Only the Tibesti, rising in the
midst of the Sahara and a counterpart of the Ahaggar,
now remains incontestably Nigritian in the hands of
the Tibbus. Only here in this remote corner, in a blind
angle away from the Libyan Desert, do we find a remnant of the ancient Saharan race which has remained
more or less intact.

In analyzing this tremendous transformation which
has taken place, we must not underestimate the influence of Islam and the Arabs; but the movement was
certainly under way before the Mohammedan drive was
launched; and it was undoubtedly the introduction of
the camel which precipitated it, for this meant the
coming—we might almost say the creation *ex nihilo*—
of the great nomadic tribes, ranging far afield, turbu-
lent, warlike and plundering.

Up to the time of its appearance, the Roman Empire was never interested in the Sahara. Its original
boundary and frontier of colonization are perfectly
known. It excluded not only the desert itself, but even
the steppe region of the high plateaus. Nor can we see
that the empire would have encountered any military
difficulties in protecting this frontier, for the Saharan
Negroes were far from being neighbors to be feared.
Byzantine Africa, however, was unable to maintain
this boundary, for it found an entirely changed situation to the south, a proximity which, it was plain, held
new dangers.

The merchants of Leptis Magna, Punic of descent
and tradition, could not possibly have foreseen, when
they imported the camel to facilitate their already

established commerce with Black Africa, how many, far-reaching and conflicting would be the results. The introduction of an animal so ideally adapted to the desert conditions was obviously of great economic importance to the country; by rendering accessible the most remote corners, it gave the first impetus to colonial expansion and resulted in the exploitation of existing springs, the drilling of wells, the introduction of irrigation and intensive agriculture, and the establishment of permanent settlements. At the same time, however, it brought with it the great nomadic whites, the Berbers from the north, and later the Arabs, less interested in developing the resources of the country than in exploiting them. By providing the perfect mount, it gave them range and power, so that they were able gradually to establish an overlordship of the desert, and become a definite menace to Roman supremacy. Thus while the camel was in some respects of great benefit to the Roman Empire in North Africa, it seems also to have been the cause of its ultimate downfall; and in this fact we have perhaps a general law. All successful colonization tends to create conditions which eventually render its continuation superfluous and impossible. This, indeed, is its aim, and in some measure its moral justification.

BIBLIOGRAPHY

FROBENIUS, LEO, *Kulturgeschichte Afrika's,* Frankfurt am Main, 1934. Excellent photographs.

FROBENIUS, LEO, and HENRI BREUIL, *Afrique,* Paris, Editions Cahiers d'Art.

GAUTIER, E.-F., *Les Siècles obscure du Maghreb,* Paris, Payot, 1927.

——"Resultats de da mission Gautier-Reygasse au Tassili," *Comptes rendus de l'Académie des Inscriptions,* 1934.

GAUTIER, E.-F., "The Monument of Tin Hinan in the Ahaggar," *Geog. Review,* July, 1934.

—— "The Ancestors of the Tuaregs," *Geog. Review,* Jan., 1935, p. 12.

GSELL, STEPHEN, *Histoire ancienne de l'Afrique du nord,* Paris, Hachette, 1913. Vol. I and following volumes.

PART IV
REGIONS OF THE SAHARA

XI
Egypt

T seems most natural, in considering the various regions of the Sahara in detail, to begin with Egypt, which is really a world unto itself. It is assuredly not only the most distinctive section in its physical aspects, but the most important from the human point of view as well. For this ancient country was the center of civilization whose influence from earliest times radiated over the entire Sahara.

We are not here concerned, however, with the study of Egypt itself. It is, of course, essentially an oasis; but by reason of its vast size, as well as its world importance, it is so exceptional in the category of oases that it manifestly emerges from this classification and becomes a separate region in its own right, as are the Maghrib and the Sudan. The only part of it which interests us in this study is the Egyptian Desert, forming the eastern extremity of the Sahara; and incidentally also the enormous importance of the influences which Egypt and the desert have had upon each other, their reciprocal reactions.

THE SEACOASTS

The most unusual feature of the Egyptian Desert is the fact that it has two important seacoasts which

give it a maritime front along its whole northern and
eastern extent. These coasts are of exceptional length,
the northern extending through 8 degrees of longitude
and the eastern through 8 degrees of latitude, giving
in all a total development of nearly 1,250 miles. It is
not in length alone, however, that these coasts are remarkable, but even more particularly in their tremendous economic importance.

For the seas that wash these shores are the Red Sea
and the Mediterranean, whose coasts were most an-
ciently populated by the human race and formed the
cradles of the two great civilizations, eastern and
western. Since the days of the Phœnician and Greek
merchant fleets the Red Sea has been the link which
joined the eastern Mediterranean with the great tropical oceans, and together these two have formed an
immense maritime route which from earliest antiquity
has not only been furrowed by an active commerce,
but has served as a lane to put the occidental world into
communication with the teeming human hives of India
and the Far East. They are the two which are now and
always have been the most important in world commerce.

In the early days, the agents of this commerce were
the Himyarites, who occupied the southwestern corner
of Arabia. This was the Pun country of the hiero-
glyphs, and these were the Phœnicians, the maritime
people of the Indian Ocean. They were in fact the real
people whose semilegendary representative was the
Queen of Sheba of sacred history; moreover it was they
who gave their name to the Red Sea, for *himyar* is the
Arabic word for red. This Himyarite commerce dates
back at least three thousand years before the Suez
Canal; and in the entire history of world commerce

there is probably no point of comparable importance
on the whole globe.

This uninterrupted flow of commerce passing for
ages through the Red Sea has washed up on its shores
a kind of residue of humanity which forms a very thin
strip along both coasts, like a plating upon the humanity of the desert, much as the banks of coral on
which they live are a plating upon the ancient rocks
of the interior. For these coast people are of a type
distinctly different from the Bedouins and desert
dwellers. They were called ichthiophagists by the ancients for the reason that they were fish eaters, a name
which they continue to merit even though they are no
longer called by it. The term in itself is an indication
that the almost nonexistent alimentary resources of a
desert soil are entirely disproportionate to the needs of
a relatively dense population congregated on its border
by the attraction of navigation and commerce. These
people belong to the sea and maintain themselves by it.

Settled, moreover, upon a soil which provides them
with almost no resources of drinkable water, this artifi-
cially massed population has covered the whole coast
with a series of cisterns which in arrangement and
maintenance form an economic masterpiece of atavistic
ingenuity. Similar cisterns are also found along the
Mediterranean coast of the Egyptian Desert, in Marmarica and around Matruh, which was the site of ancient Parœtonium. Cisterns such as these are quite unknown in the rest of the Sahara; and their presence in
these particular localities bears witness to two facts:
that in order to provide for the exigencies of a great
commercial navigation route it has been necessary to
violate nature and create life in the desert; and that
these people, thanks to the intellectual and pecuniary

resources their traffic brought them, have been enabled to accomplish this.

Another striking feature of the influence of this far-reaching commerce is the amazing architectural prodigality to be found in the Red Sea ports, not alone on the Asiatic coast where the Mecca pilgrimages have kept them flourishing, but along certain parts of the African coast as well, as for instance at Suakin. The houses are handsome with balconies, verandahs and *mushrabias,* or lattices, of beautifully carved and fretted woods which make an extraordinary contrast with the meagerness of the native vegetation. The woods thus used have been imported by sea from distant countries, especially from Java; and this one fact in itself is most illuminating, for it shows not only how far this commerce has extended, but the prosperity it has brought in its wake. [See illustration facing this page.]

The Egyptian coasts of the Red Sea and Marmarica in ancient times were thickly strewn with such celebrated ports as Berenice, Leuke Kome, Myos Hormos and Parœtonium; but practically all of these have now disappeared. Even the few which still remain, such as Kosseir, the Leuke Kome of antiquity, and Matruh which was formerly Parœtonium, are but poor substitutes for these earlier ones. This is because steam and increasing tonnages have killed the small ports of call; but since, on the other hand, these are the factors that have been responsible for the opening of the Suez Canal, the compensation has been more than ample.

The coastal advantage that Egypt thus enjoys is not shared by any other portion of the Sahara. It is true that the opposite extremity of the desert has a very extensive oceanic coast line, but this coast faces toward distant America; it has not now and never has had any

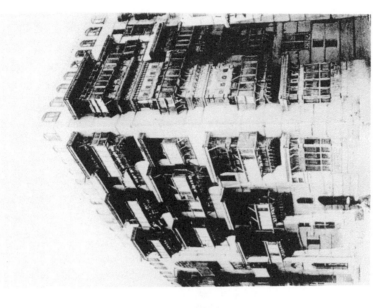

Photograph by P. Savignac

FRETTED WOODWORK IN THE SACRED CITIES OF JEDDA AND YAMBO

commercial relations with the rest of the globe. As far as its importance in human affairs is concerned, it is practically nonexistent. The great mass of the desert there lies shut within itself, completely alien to the outside world. Of the whole Sahara, Egypt alone is specially favored by reason of its coasts.

THE NILE ROUTE

Another exceptional feature of the Egyptian Desert is of course the Nile. This splendid river has been the fertilizing factor in the actual creating of the country; it has also served as a great living waterway by which the tropical aquatic fauna have been enabled to cross the desert and reach the delta, so that crocodiles and hippopotami would still be Mediterranean denizens had man permitted it. But more than this, it too has formed a great open route of traffic and communication, one that crosses the entire width of the desert, making Egypt exempt from the trans-Saharan problem which is elsewhere so grave. The problem here has been solved by nature itself.

By means of the Nile, Egypt has always been in open communication with Negro Africa: with Nubia, which as far as Merowe is strewn with Egyptian monuments; and more particularly with Abyssinia, which was the ancient kingdom of Axum. Our scholastic texts, dwelling too exclusively on the occidental phases in their exposition of history, do not sufficiently emphasize this fact, which was of tremendous importance.

For the Roman Empire, establishing what today we would call a protectorate over the kingdom of Axum, converted it to Christianity; and, as we would also consider it today, brought it civilization. Abyssinia, directed, financed and given the support of the Roman fleet in the Red Sea, was thus put in a position to conquer and wipe out the Himyarite kingdom of southwestern Arabia; and in this way the great maritime route to India and the Far East was opened without opposition to the commercial enterprises of the mercantile Greeks of Roman Egypt—a triumph of financial imperialism which in the long run was to prove very costly. For Arabia, cut off from its ancient source of prosperity, the sea, was driven perforce through economic strictures to seek new projects. The result was the explosion of Islam, which rocked the entire world. It is a curious fact, however, that in thirteen centuries the tide of Mohammedanism has never been able to engulf Ethiopia, which has always remained Christian, so deeply did the seed sown by Roman Egypt take root.

It was also by way of the Nile that exploration during the nineteenth century was first enabled to penetrate successfully into the interior of Africa; and one of the first problems to be raised and solved in regard to Central Africa was that of the sources of the Nile. Today the whole Saharan portion of the Nile Valley is open to tourists, for a regular steamboat service goes as far up the river as the second cataract at Wadi Halfa, making one stop at Aswan on the first cataract. There is also the railroad which goes as far as Khartum. in the heart of the Sudan, at the juncture of the Blue Nile and the White Nile, where another railroad makes direct connection with Suakin on the Red Sea. Thus the problem of a trans-Saharan railroad is here adequately and easily solved by the help of the river. This is the first step toward what is the dream of the future, a transcontinental railroad from Cairo to the Cape.

The Egyptian Desert: Influence of Its Arrangement on Human Distribution

The combination of these great routes of maritime and fluvial navigation, together with the peculiarly elongated shape of the Egyptian Desert itself, have imposed upon this section of the Sahara a distinct and unique organization of human life. We must remember that this desert forms a narrow corridor some 870 miles long with an average width of not over 250 miles. To circulate within his own confines, the Egyptian has not only the great river that bisects this corridor lengthwise, but over and above this the two bordering seas. In short, there is no corner of the Egyptian Desert so remote as to be more than 125 miles from either a maritime or a river base.

Throughout the entire duration of Egypt this natural plan of the desert has facilitated its surveillance and complete domination by the masters of the valley, since the days when such authority was represented by the chariot and infantry forces of the Pharaohs or those of the Roman Legion. In our own times the narrowness of the Egyptian Desert was very helpful to the automobile squadrons of the English Army during the Great War, by sufficiently restricting the field of campaign about the provisioning bases; it also permitted the infantry to render useful service in spite of its dependence on haphazard railway facilities. At present it is the proximity of the maritime base at Suakin which permits England to maintain her domination over the Egyptian Sudan even though she has renounced it, nominally at least, over Egypt itself.

A glance at the map gives at first only a rather inexact idea of the actual situation of this Egyptian

Desert. We see plainly that it is bounded on the east and north by the Red Sea and the Mediterranean; but it takes some reflection to perceive that it is no less bounded on its western side by the solitudes of the Libyan Desert and the Erg. Such an erg, the most gigantic and uninhabited on the face of the earth, is both an obstacle and a protection, in each respect as efficacious as the sea and much more difficult to pass. Consequently any land communications of Egypt with the outside world are practically closed except to the south, in the direction of Black Africa, which is not a dangerous neighbor. There is no contact with formidable Asia except by way of the Isthmus of Suez; while in the direction of North Africa, the only communication with the restless and warlike Berber world is through one narrow passage between the erg and the Mediterranean, a strip of land which in function is almost as much an isthmus as the other. It is guarded by the famous oasis of Jupiter Ammon, or Siwah as it is known today; and the continued renown of this name throughout history is enough in itself to show the importance of this single outpost of communication between the two worlds.

In many respects the Isthmus of Suez and the Isthmus of Siwah, if we may be permitted to call it so, are counterparts. Throughout the centuries they have not only blocked foreign invasion, but inversely they have checked any imperialistic expansion of Egypt, keeping her confined to her own boundaries, as isolated and untouched as if under a bell jar. The Egyptian Desert is a kind of peninsula. When we combine this fact with the ease of communication within the territory itself, which has contributed so largely to its control, it is evident that we have the explanation of the immense duration and stability of the Egyptian Empire and its unity throughout thousands of years, as well as the gradual and progressive evolution of its institutions, which never suffered any radical interruption either from outside invasion or from internal disorder.

UNIMPORTANCE OF THE NOMADS

There is another factor in the physical constitution of the Egyptian Desert which has had a considerable effect on the social organization of its population. This is one to which we have already referred briefly in our study of the Quaternary erosion, the curious asymmetry of the two desert belts which are separated by the Nile Valley and which constitute the two halves of the Egyptian Desert. The eastern half, known as the Arabian Desert, is mountainous and scored by dead wadis whose fossil networks are still defined with perfect clarity. On this side is what we have called a very young desert relief, marking the beginning of a cycle.

The western half, which is really a portion of the Libyan Desert extending from the left bank of the Nile to the first dunes of the Great Erg, is almost exactly the reverse, being an immense uniform plateau of very moderate altitude. In this region it is impossible to find even the general outlines of any fluvial system whatever. Those which must necessarily have been engraved on the face of the desert by the rivers of the Tertiary epoch have been exposed to the desert climate for so immeasurably long a time as to have become confused and obliterated. Such traces as doubtless still exist can no longer be deciphered, and we find in the Libyan Desert what we have termed a senescent desert relief.

If we seem to insist once more on a fact that has already been fully discussed, it is because it is of considerable human importance and has a particular bearing in this connection. We have shown how the useful effect of the superficial waters is intensified by the natural drainage of the Quaternary wadi systems, and that this has a close relationship with the distribution of watering places and pasturages, and in consequence with the life of the pastoral nomads. Judging by our observations of the Sahara as a whole, and making of course such reservations as we should in attempting to reduce the complexity of natural phenomena to an exclusive formula, it seems permissible to state as a general rule that nomadic life is lacking where there is no young desert relief, which is to say no networks of fossil wadis.

We should expect, then, to find a noticeable absence of nomads in the Egyptian Desert, for all the conditions are particularly unfavorable to this type of life. In the first place its attenuated shape is already sufficiently adverse to arrest any development on a large scale, since the great nomad tribes have need of immense spaces for their activities; and this circumstance undoubtedly is greatly aggravated by the senescent relief of the Libyan half of the desert. At any rate, the supposition is certainly confirmed by the facts, for the Egyptian Desert has no nomads. Or at least, to be somewhat more exact, it has none of the fierce, powerful nomadic tribes of maurauding outlaws, warlike by instinct and inclination, such as elsewhere constitute a constant menace to law and order unless a community is resigned to commit the public welfare to their charge. Egypt, in the way of nomads, has only its Bedouins.

The word Bedouin, from an old Arabic word "Bedaui," is one that was naturalized into all the European languages at some indeterminate date. It may possibly have come into use through the Bonaparte expedition into Egypt, for the connotations which we give the word correspond to the Egyptian usage. The term evokes for us the idea of a subservient humanity, a furtive human vermin, an army in retreat; the belittling sense of the word is undeniable. Curiously enough the French word Bédouin is practically obsolete in Algeria, where it could hardly without gross misrepresentation be applied to the Algerian nomads, who are "Sons of the Great Tent" and represent a true indigenous aristocracy. But in Egypt the word Bedouin, with all its current implications, does accord with the reality; for the Egyptian Bedouins are hardly more than humble dragomen and guides for tourists about the pyramids. Even in the Arabian Desert, where they are in their natural habitat, they are just a peaceful caravan people who circulate between the Nile and the Red Sea. In Marmarica, however, where they are in free communication with the Sahara of the great nomadic Arab and Berber tribes to the west, we find the Uled-Ali Bedouins reflecting these outside contacts by supplementing their peaceful profession of caravan convoy with that of running contraband.

The absence of a strong nomadic population in Egypt is emphasized by the very tardy introduction of the camel. It was never adopted during the time of the Pharaohs, in spite of the proximity of Asia and the easy relations with Arabia, where the dromedary originated. It required nothing less than a foreign invasion, the Persian conquest, to bring the camel to the banks of the Nile. And not even then nor ever since has the camel seemed to make himself really at home in the Egyptian Desert. During the Great War the animals for the British corps of meharists had to be recruited from all the deserts from Tunis to Bombay, for the resources of Egypt itself as regards meharis or mounts were entirely insufficient, almost to say nil. Even the model of the saddle had to be imported from the Sudan, for Egypt has no proper riding saddle, the only ones in use being packsaddles and makeshifts. This backwardness in the use of the camel is in itself a clear indication of the subordinate position of the nomads, who never attain greatness without a domesticated camel. [See illustration, p. 124.]

The whole political and social life of the Orient, from Central Asia to the Atlantic coast of the Sahara, is based upon the equilibrium between the two great divisions of humanity, the nomadic and the sedentary peoples. Between these two elements, so dissimilar in modes of living and oftentimes in race, there is daily collaboration; it is a combination in which harmony can be established only by the subordination of one or the other. In Egypt, throughout thousands of years, the sedentary folk have always held the nomads at their mercy. This has been not only because Egypt as a gigantic oasis has nourished an enormous population of sedentary people (given by a recent census as some fifteen millions) whose power has lain in their grouping and organization; but equally because the desert of Egypt, constituted as it is, could not maintain any powerful nomadic tribes whose energy might have made up for a deficiency in numbers.

In this connection we might consider Chaldea, which represents the opposite pole of oriental civilization. As an oasis Chaldea cedes nothing to Egypt; the essential difference lies in the fact that it is surrounded on all sides by "living" deserts swarming with nomads, in direct contrast with the "sterile" deserts, if we may call them so, of Egypt. The mountains and high plateaus of Assyria, Media and Persia and the desert wadis in

Syria and northern Arabia have always been an inexhaustible reservoir of powerful nomadic tribes. The
Assyrian armies were the terror of the oriental world;
and this whole world was twice unified into a single
great Chaldean empire, once under the Persians, and
a second time even more completely under the Arabs.
In fact the great Mesopotamian oasis has been dominated by each of these numerous tribes in turn, but
seems never to have had an identity of its own. Chaldea
under nomadic domination has had a great military history, but one which is as heterogeneous and interrupted
as that of Egypt is unified.

The whole history of Egypt is one of nonaggression
and shows the imprint of the fact that the Pharaoh was
always the ruler of a sedentary people. Even in the time
of Sesostris, Egypt was never able to accomplish anything either complete or lasting in the way of empire;
while Egyptologists have discovered a manuscript
dating from the time of Rameses which has been beautifully translated by Maspero and proves to be what we
would now call an antimilitaristic pamphlet. The sentiment expressed in it would seem to be that of Egypt
of all times. Even today it is certainly not the Egyptian
who is the sword of Islam, but the Turk.

Civilization, on the other hand, would seem to owe
more to Egypt than to Chaldea. The oriental historians,
particularly Ibn Khaldun, have shown us that arts, industry and learning have always been the attributes of
the sedentary people of the Orient; and the more we
learn of Egyptian civilization the more clearly can we
trace in it the roots of our own. The intellectual metropolis of Islam today is Cairo; and Egypt, so ingrowing
and so individualized, seems predestined to become the
first oriental country to develop what we of the Oc-

cident call a nationalistic sentiment. All of this, of course, is the outcome of the form and peculiarities of the Egyptian Desert, and directly dependent upon them.

The Egyptian Oases

A particular feature of the Egyptian portion of the Libyan Desert is the four groups of oases, Kharga, Dakhla, Farafra and Baharia, all of which are of very ancient fame. They were, in fact, the ones to which Herodotus, giving a Greek form to an Egyptian word, originally applied this happy term "oasis." Each of them, indeed, is the typical oasis, the single fertile spot lost in the immensity of the desert like an atoll in the vastness of the Pacific. No line of verdure links them either with each other or with the valley of the Nile. They bear no relation to any wadi whatsoever, nor to any superficial circulation of the pluvial waters. Each owes its existence to a local emergence of the deep-seated aqueous layers.

They have all been thoroughly investigated and the Geological Survey of Egypt has published a splendid monograph on each of them. A glance at the geological map is sufficient to enlighten us as to their true nature. Each oasis coincides with an outcropping of the older and deeper geological layers, through some break in the outer covering of more recent strata. Kharga and Dakhla mark the geological contact between the Cretaceous limestones and the Nubian sandstones. Farafra is at the point of contact of the clays and sandstones of the Upper Cretaceous with the limestones of the Lower Eocene; while Baharia is an anticlinal buttonhole through which the underlying Nubian sandstone emerges, rupturing the deep Lower Eo-Cretaceous and Eocene beds. At each point the Geological Survey has

Service photographique du Gouvernement Général

ARTESIAN WELL, OASIS OF TOLGA, NEAR BISKRA

established the existence of folds, joints and other more
or less active earth movements, sometimes accompanied
by eruptive occurrences.

The subterranean waters thus brought to the surface
emerge in a thermal state. At Kharga and Dakhla they
are really hot, especially at Dakhla, where they have a
temperature of 102° F. The natives, who are perhaps not
entirely trustworthy, claim to remember a time when
they were hot enough to cook eggs. At Farafra and Baharia the temperature is not so high but is accompanied
by a discharge of gas strong enough to blow the cork
out of a bottle.

In the vicinity of Kharga and Dakhla the water
sometimes gushes from artesian springs whose outlet is
the craterlike center of a small clay knoll, a form which
first captured the attention of Herodotus. More often,
however, in order to utilize these resources of water
man has had to employ artificial means. Especially at
Dakhla it has been necessary to dig wells to reach the
water layer, an idea evidently derived from the natural
springs themselves. As we go farther north there is no
longer pressure enough to cause the water to gush; at
Baharia and even at Farafra it simply flows from ordinary springs, and here again man has resorted to his
ingenuity to exploit the existing supply. In this case he
has done so by means of long underground aqueducts,
of exactly the same type as those called *foggaras* in the
Algerian Sahara. In fact these two regions are the principal ones where considerable subterranean labor has
been required to insure the maximum utility of the
available resources; and the similarity of the methods
and techniques involved forms an interesting comparison, as we shall later see.

The artesian wells and foggaras of this region are

real works of art. The wells must be sunk to a depth of several tens of yards, and are solidly built of acacia wood; while the foggaras are constructed of stone in walls of excellent masonry. By the time the Europeans arrived, according to notes of the Geological Survey, they found the native well makers equipped with a stock of tools which already showed decided evolution, including a long metallic rod for boring through the last hard stratum. Since Egypt was always the center of progress and civilization in the Orient, it is natural that methods here should have been more advanced than elsewhere.

The original technique however is assuredly very ancient, for the inhabitants of the oases were already celebrated well makers at the time of Olympiodorus in the sixth century A.D. On the other hand, it cannot be much earlier than this, for both wells and foggaras would seem to date definitely from the Roman period. The old wells in fact are called "romans" by the natives, although this in itself would not be convincing proof. But archeologists claim they are right, and also that the very fine workmanship of the foggaras of Baharia is certainly of Roman technique, indicating a civilized administration as well. Moreover, Farafra, prior to the Roman period, must have been the "cattle country" of the ancient Egyptians, which it has certainly ceased to be. Cattle grazing does not seem to be compatible with intensive agriculture of the type made possible by the shelter of the palm trees, and must have gone out when the wells were introduced. It appears then that the Roman Empire was responsible for the systematic exploitation of the artesian waters and the foundation of the palm groves with their attendant cultivation, as well as for the propagation of the domesticated camel. The

two things in fact are closely associated, since the remote corners of the desert cannot be made of any value
unless rendered accessible. The problem of communications is inseparable from that of agriculture. All of this,
which accords perfectly with data of the same order in
Algerian territory, indicates that the Sahara owes much
more to the Roman Empire than we might be inclined
to think.

In actual numbers the Libyan oases have only a very
few inhabitants; Dakhla has about 17,000, Kharga
about 8,000, Baharia 6,000 and Farafra only 632 in all.
But their importance is not to be measured by their
numerical population; in this desert, with its aging
surface relief, these oases are the only watering places.
This fact again reflects the absence of any strong nomadic population. For this single chain or system of
oases marks the one open route in existence by which
the length of the Libyan Desert may be traversed; the
only one, that is, aside from the Nile Valley itself, and
which is not only distant from it but independent of it.
The importance of this route is further enhanced by the
fact that the place where it joins the Nile, at Abydos,
near Thebes, is also the exact spot where the great
routes of the Arabian Desert, coming from Kosseir,
Myos Hormos and even Berenice on the east coast, converge on the opposite bank.

This crossing of the desert roads has made Upper
Egypt very distinct from Lower Egypt. The environs
of Abydos are the foremost site of Egyptian prehistory;
ancient tombs are found there antedating even the first
dynasty, and it was there that Thebes of the Hundred
Gateways rose, the historical rival of Memphis. The
particular god of Thebes during the time of the
Pharaohs was Ammon Ra, the god with the ram's head,

identical with the Jupiter Ammon of the Romans; and
it is interesting to note that this god was the only one
in the whole Egyptian pantheon whose cult, by way of
the oasis route, was disseminated through the Sahara.
In fact the Oasis of Siwah, which marks the opposite
end of this route on the edge of the Maghribi Desert,
was chiefly celebrated for its temple of Jupiter Ammon
and bore the name of that god throughout antiquity.

The oasis route was always of tremendous military
importance. There is a legend recounted by Ammianus
Marcellinus according to which Thebes would seem once
to have been captured by the Carthaginians; but naturally it must be kept in mind that some memory of
Berber raids, sweeping unexpectedly out of the desert,
may have figured confusedly in this tale. At any rate, it
was by way of this route that the army of Cambyses set
out from Thebes in its vain attempt to conquer the
Occidental Desert, and the name Kharga signifies the
"sortie gate." The Persians, with their atavistic nomadic instincts, were much interested in this "sortie
gate" when they were masters of Egypt; and it was at
Kharga that Darius erected the lovely temple of Ibis.
[See illustration facing this page.]

The oasis route again came into strategic value during the Great War, when it drew the Senusi attack
on Egypt. The Senusi, trickling in by way of Baharia and Farafra, captured Dakhla; and for a long
time the front lay between Dakhla on one side and
Kharga, held by a detachment of the English Army, on
the other. At this time the pretty, luxurious little railway, built in peace times for the tourist trade, served to
provision the defenders.

Photograph by Gautier

COLONNADE OF THE TEMPLE OF IBIS AT KHARGA (VIEWED FROM ABOVE)

The Isthmus of Siwah

Going northward, the oasis route comes out at Siwah, which is the end of Egypt and the beginning of another Sahara. It is the narrow sand-free passage which puts into communication two separate portions of the desert otherwise practically sealed from each other; it is the only link between the Shark and the Gharb, the east and west halves of the Moslem world which differ so profoundly, as all the Arabian writings show. Of the inhabitants of Siwah very little is known, but at least one fact is certain. They speak neither Arabic nor Coptic, but a Berber dialect. In this particular we are upon the threshold of the Barbarian world, the Gharb.

We have as yet only a very general idea of the Siwah passage or "isthmus," without any detailed topographical or geological information. The oasis lies at the base of the great escarpment which forms the southern boundary of Marmarica, a very long continuous cliff which extends from Siwah to the delta of the Nile. This steep cliff is the line of geological contact between the Miocene beds of Marmarica and the adjoining Eocene beds; apparently the Siwah springs, like those of the Egyptian oases, are fed by the deep-seated aqueous layers. The base of the escarpment is elsewhere marked by infrequent outlets, which, however, do not feed any oasis although they do in some places lead to the formation of shotts. The best known of these is the Wadi Natrun, at the eastern extremity adjoining the delta, where the native carbonate of soda is exploited. Another of these shotts is that of Moghara, which has a good watering place and is just about halfway between Siwah and Cairo.

It is Siwah itself, the oasis of Jupiter Ammon,

which has had the great name and was the famous historical center. It was there that the Pharaohs went to have their auguries read on coming into power; and their successor, Alexander the Great, made it his first duty to go there and have himself anointed Son of Ammon by the High Priest. Its continued importance lies in the fact that it is the junction, the knot that ties together all the desert routes, and in this respect is the real landward key to Egypt.

It is utilized by two east-west routes, one of which passes directly through Siwah and crosses the width of the Libyan Desert, parallel to the coast. This, however, is a desert route which is difficult to follow and consequently to give an account of, except that it goes by way of Moghara and comes out at the capital of Lower Egypt.

The one which is most frequented, and which the Arabian geographers never fail to describe, is that which follows the sea. It is the coast route of Marmarica, where the Uled-Ali Bedouins lead a precarious existence by caravan and smuggling, and this thin cordon of humanity along the edge of the coast forms the real link between the East and the West. The route is actually as much a maritime as a land route, for the infrequent watering places have always furnished fresh water and shelter for the coastwise trade. In fact the Shark and the Gharb are so completely separated by the Libyan Desert that, properly speaking, they communicate by sea, for it is difficult to state the exact degree of mutual assistance given by the caravans and vessels to one another along this semimaritime route. It was used some two thousand years prior to the Arabian conquest for the oriental colonization of the Phœnicians and Carthaginians; and many armies have followed it since the coming of Islam, including the successive Arabian armies going out to the conquest of the Maghrib, and the Fatimid armies advancing on Egypt from the west. It is clear that all of these must have depended on a supporting fleet for their provisions.

Nevertheless, the only port of any importance on all this stretch of coast was Parœtonium, which we know today as Matruh; and this one was important only because it was the nearest to Siwah and made the most direct connection with it. It has always been Siwah that was the true center of communication; and it is for this reason that it seems permissible for us to call the whole strip of territory by the same name, the Isthmus of Siwah.

BIBLIOGRAPHY

Explorations of Prince Kemal ed Din.

FOURTAU, R., "La Marmarique," *Bulle. Soc. khédiviale de géog.,* Cairo, 1907.

GAUTIER, E.-F., Articles in *Annales de Géographie,* Vol. XXVII; *Revue de Paris,* Jan. 1, 1918.

Geological Survey of Egypt, various publications already cited.

HUME, W. F., *Geology of the Eastern Desert,* Cairo, 1907.

SCHWEINFURTH, G., *Aufnahmen in der östlichen Wüste,* Berlin, 1900, 1902.

—— *Auf unbetretenen Wegen,* Hamburg, 1922.

XII
The Tibbu Sahara

THE precipitous cliff of Siwah is continued westward by the one which forms the southern limit of Cyrenaica. It is the same geological boundary, and it too is marked by oases. One of these is Jerabub, which with Kufara is joint-capital of Senusism; and another Aujila, whose palm groves were already known and discoursed upon at length by Herodotus. At Aujila we are approaching the vicinity of the great Syrtis of the Tripolitan coast, and the main desert route from Egypt to the Maghrib passes through it. But southward from this route and coming into immediate contact with it extends the worst and most awesome part of the Libyan Desert, those dreadful solitudes across which Aujila and Jerabub communicate so painfully with Kufara. It is always the sealed compartment; and within it, as in a blind angle, protected and removed from the general traffic, we find the Sahara of the Tibbus. These are the only black Saharans who have retained an independent racial and political existence right up to the present day.

Among the results of their isolation must be counted the difficulties of exploration. Barth and Rohlfs tried in vain to penetrate into the Tibbu Sahara, and Nachtigall was the first European who succeeded in doing so. For forty years his book was the only document on this region, and to a certain extent still remains so. The whole country has now been occupied by the French troops of the Sudan, and some rather scanty data may be found in the *Renseignements coloniaux* published by the Comité de l'Afrique Française. Even more important material on this region has been furnished by the Tilho expedition of 1912-17, which made a long study of it, and the Tilho map particularly gives us a perfectly clear image of the Tibbu Sahara. This has also been supplemented by information on the Tibesti massif brought back by the Dalloui expedition. But while these later French studies are naturally more precise and scientific than the Nachtigall account, of which incidentally they confirm the authenticity, nevertheless Nachtigall's description still remains indispensable for an intelligent idea of the subject.

THE TIBESTI

The Tibbus have one great stronghold which has remained impregnable throughout all the ages. This is the Tibesti, a counterpart, a twin of the Ahaggar. Together these two mountainous masses of analogous structure dominate the desert. The Tibesti is a gigantic massif, roughly triangular in shape, each side of the triangle having a development of from 250 to 300 miles in a straight line. Amid the surrounding depressions whose altitudes nowhere exceed a few hundred feet, the abrupt upthrust of this lofty prominence is extremely arresting.

It is of volcanic origin, made up of eruptive rocks that have emerged through a basement of crystalline rocks covered by great thicknesses of horizontal sedimentary layers, in which the fossiliferous Silurian period is represented by sandstones. From measurements which seem to be highly guaranteed for accuracy, the Tibesti has the highest summit within the entire Saharan area. This is Emi Kusi, a superb volcano with an altitude of over 11,000 feet. Tilho has given us a detailed map of it, showing it to be perfectly intact, crowned by an extinct crater with fumeroles, or vapor vents, and comparable in dimensions and general plan to Etna. Nor is this recent, practically active volcano by any means the only one in this range. Tilho saw and sketched several other volcanic cones; and even Nachtigall saw a crater which was very regular in form and lined with natron. Moreover, the "Thundering Spring" of Soboro, famous among the Tibbus, is a sulphurous spring where the water at a temperature of 185° Fahrenheit boils with explosions. [See figure, p. 163.]

The Tibesti, like the Ahaggar, was once a watershed for some of the great defunct wadis, and every face of the massif is deeply scarred with a compact network of their valleys. Moreover, geologists have found laterite in this region which could only have been formed in a relatively humid climate from the subfossil skeletons of elephants; while in one of the Tibesti water holes the Tilho expedition found a living crocodile, survivor of the residual fauna and a brother of the Tuareg crocodile. In the Tibesti, as in all the mountains of the Sahara, the fauna as well as the surface relief show evidences of a more humid climate in the Quaternary period, the one immediately antedating our own.

The Tibesti today, however, belongs decisively to the Saharan domain. It is prolonged to the southeast, after an interruption, by the plateaus of Ennedi, which have an altitude of about 4,000 feet, but these are

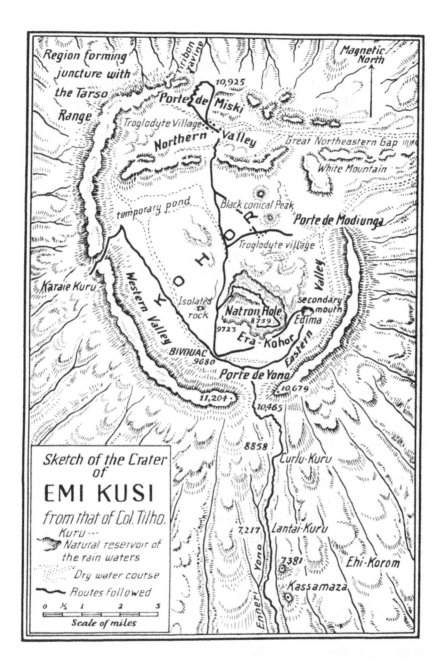

EMI KUSI

already merging into the Sudanese steppe. Nevertheless the Tibesti itself, in spite of its great elevation, is definitely of the desert. It presents a lunar landscape of denuded rocks; and the too-infrequent rains which are arrested by it in passing, only very rarely, in certain favored sections, succeed in feeding some stream of running water, and even this will be charged with natron.

It is in this setting that the Tibbus live. The Tilho expedition estimated their numbers at about 10,000 in all; and these first evaluations, made at sight, have proved in every country of the world to be highly exaggerated when compared with those ultimately furnished by a regular census. In any case, they form at best but a very meager population. As a people, however, they are infinitely interesting by reason of the fact that they themselves are a kind of residual fauna. By the color of their skin and the affinity of their language they belong to Black Africa, being related to the group which includes the Bornuans and Kanuri, their neighbors to the south. Also, among their traditional weapons they have the throwing knife, which is Central African. Yet they are not true Negroes; they do not have the typical characteristics of woolly hair, thick lips and flat nose. At the same time, they are not as varied as is a mixed population; their homogeneity was one of the things that caught the attention of Nachtigall. They are a definite race, with a clearly defined type.

They are noted for a marked sobriety and the faculty of enduring extraordinary privations; their chief physical characteristic is an extreme leanness with tenuous muscles, which does not mitigate against great vigor and more especially endurance; above all they are known among their neighbors for a more-than-human,

Photograph by Désiré

A CANYON OF MUIDIR (GORGES OF TAKUMBARET) IN THE SILURIAN
(POSSIBLY CAMBRIAN) SANDSTONES

almost animal-like agility. This agility in fact is the essential ethnic characteristic of the Tibbus. Captain Ballif in this connection relates the following story:

In 1912 one Tibbu, after a raid in Aïr and the death of all his comrades, set out alone and on foot for the Tibesti, more than 600 miles away. He carried with him for subsistence only the raw meat of one goat that he had killed and as much provision of water as the goat skin could contain. . . . He came through wonderfully well in spite of the hardships of such a journey.

It is highly improbable that either a Berber or an Arab would be capable of such an exploit.

Nachtigall, in considering the identity of an ancient Ethiopian tribe mentioned by Herodotus in the vicinity of Fezzan, believed that they were these same Tibbus, simply because they were described as "the most agile of men." The Tibbus of today are quite evidently black Saharans, fashioned and physically modified by the desert; perhaps we may be allowed to presuppose a desert other than the present one in which the camel had not yet appeared and the problem of getting from place to place imposed lasting modifications upon the human organism. They would certainly seem to represent the last descendants of the Saharan Ethiopian whose domain, according to the ancients, once extended all the way to the foot of the Atlas; and it is permissible to picture this early ancestor in their present image.

BORKU

The Tibbus are not exclusively a mountain people dwelling in the Tibesti. They inhabit also Borku, the immense very low basin which lies between the massif and Lake Chad. The various sections of it include Borku itself, Bodele, Egei and Bahr-el-Ghazal, names which have all had a certain renown because of the controversy raised by Nachtigall and later settled by Tilho. This basin is the common alluvial region of the wadis coming down from the Tibesti on the north, and up from the Shari to the south. It is what Tilho has called the Chad Netherlands.

These low lands have a mean elevation of only about 650 feet above sea level, which is notably below that of the lake itself, whose level is over 800 feet. It is perfectly certain that this whole immense basin, of such indeterminate and uneven level, once and quite recently contained a vast marshy lake, for Nachtigall found here innumerable fish skeletons and mollusk shells which later were carefuly examined by the Tilho expedition. There is nothing in any way unnatural about this; similar phenomena are observed, as we have shown, in the depressions of all the closed desert basins.

All that is left now of this original lake is Chad, the residue and present representative of its former greatness. But through the light, more or less permeable and often sandy earth which lines the bowl, Chad has an outlet and a subterranean flow; it is really a visible portion, a sort of aneurism or breaking-out, of a great sheet of water which lies beneath the surface and spreads its ramifications under the entire "Netherlands." While the whole depression is desertlike as regards climate, including the slight hogback which skirts the north shore of the lake, nevertheless the aqueous layer is almost everywhere present within the depths of the soil. It is easily accessible in all the hollows, where it feeds camel pasturages or oases. The oases are particularly well developed in Borku, and excellent dates are grown which were specially praised by Nachtigall, who made a stay at Aïn Galaka. Another of the Borku oases is

Faya, headquarters of the French occupation and base of operations for the Tilho expedition.

This entire basin which extends from the Tibesti massif to Lake Chad is Tibbu territory. The desert Tibbus are not exactly the same physical type as the mountain dwellers, being less clearly individualized and of less pure race, more Negroid. But they are Tibbus nevertheless, and speak the same tongue. Naturally this indigenous desert population is very sparse; Nachtigall estimated it at some 10,000, and even this is probably too high. Moreover, a new ethnological element has made its appearance in Borku since an Arab tribe from the north invaded the territory sometime around the beginning of the nineteenth century and remained to settle there.

These are the Uled Sliman, originally from the Gulf of Sidra, which they left about 1820, following some trouble with the Turkish authorities. Nachtigall has given us an account of the extraordinary epic of this tribe, whose number he estimated as comprising no more than 500 foot soldiers and 500 meharists, or camel troops. With this tiny band they accomplished miracles during three-quarters of a century of uninterrupted warfare. They met with crushing defeats which threatened them with complete annihilation, but from which they always recovered with an indomitable vitality and energy. They spread terror on all sides and brought upon themselves an impotent hatred, but never lost control of the situation. It is interesting to have so concrete and contemporaneous an example to illustrate the rôle of the great white nomad who in 1,500 years has so changed the face of the Sahara.

Kufara

While the last remnant of the Saharan blacks is thus
under fire to the south, a parallel offensive, although
somewhat different in method of operation, is progress-
ing against them to the north. The scene of this conquest
centers about the oases of Kufara. We know for a certainty that Kufara until a fairly recent date was Tibbu
territory, and according to Lapierre there are ruins of
Tibbu villages not only at Kufara, but even as far west
as Wau el Kebir. The conquest of the region by the
Senusi is a recent one; but these Arabs of Kufara are
an altogether different stripe of men from the Uled
Sliman. They are a commercial people, cultured, re-
fined and intellectual—all of which in the Orient seems
to be quite compatible with a desperate religious fa-
naticism. In any case, the military conquest here has
been followed by a prolonged peaceful penetration; and
their influence is already spreading through Borku and
even into the Tibesti, for they push ever farther in
pursuit of their commerce, and as they go they build
their mosques, acquiring proselytes along with their
customers.

Kufara is extraordinarily isolated; its situation is
without parallel in the whole Sahara. It occupies almost
the exact mathematical center of the Libyan Desert.
Setting out from it toward any point of the compass, it
is necessary to cross from 250 to 300 miles of empty
space in order to reach an inhabited locality. The whole
region around Kufara within a radius of some 300 miles
is practically untouched by exploration, and forms by
far the most extensive blank on the map of the Sahara.
It is a region which has always presented inordinate
difficulties to European exploration.

During the whole of the nineteenth century the only European who was able to penetrate it was Rohlfs, who was the first even to have seen this group of oases. Rosita Forbes in 1920-21 followed his journey with certain variations, and has given us a lively description of Kufara, while Hassanein Bey has since retraced her route and completed her data. The Great War brought Sergeant Major Lapierre into the region, and his narrative, though brief, has been very helpful, particularly as his journey in from Fezzan has revealed the existence of one route from the west which is certainly more accessible than that of Aujila. It has a certain number of watering places, two of which are especially fine. These are Wau el Kebir and Wau en Namus. Lapierre was particularly impressed with the "three great lakes" of Wau en Namus, lying in a large basin over 900 feet deep.

Kufara since 1931 has been occupied by the Italian troops, and is now connected with the Mediterranean by an excellent automobile road. A map of the region to the scale of 1 : 100,000 has also been made, and was exhibited in the Italian section of the Saharan Exposition at the Trocadéro in 1934. Consequently we have already a general idea of the locality. Kufara is set in a landscape of cliffs and isolated rocks carved by erosion out of the sandstone, a striking contrast after the endless and hopelessly monotonous plain of the serir which must be crossed in reaching it from the north. Immediately to the south of Kufara Lapierre sighted a range of mountains which might be, according to native information, a last spur of the Tibesti massif.

It is the extreme isolation of Kufara that makes it so important. It is the only stopping place on the great caravan route which puts the Mediterranean into communication with Borku, and beyond Borku with Wadai. This route, though very difficult, is greatly traveled; yet the southern portion of it, between Kufara and Borku, is hardly known at all to Europeans. The region that lies north of the Tibesti Mountains is now under exploration by the Italians; and it would seem evident that the valleys of the massif must be more or less prolonged in this direction, feeding some underground reservoir. For there is abundant surface water at Kufara which spreads out in marshes and small lakes, and flows freely without irrigation works. Had there been any artesian wells and foggaras, Lapierre during a sojourn of some months would have noticed them, especially after coming from the Algerian Sahara where they are so prevalent. And in this point lies the great mystery: What can be the source of this abundant supply of water? The Italian explorations will answer this question. In any case, this very important group of oases should, by reason of its size, support a population of several thousands.

There are certain other oases which as regards their inhabitants should be considered Tibbu territory. Among these are Gatron in southern Fezzan, and Kawar, which lies west of the Tibesti and includes the salt mines of Bilma. These, however, although racially affiliated with the region we have been considering, are located in a different section of the Sahara, which by reason of its physical make-up constitutes a separate and distinctive region. They must therefore be considered as belonging geographically to the latter, which is to be studied later under the heading of Fezzan.

BIBLIOGRAPHY

Comité de l'Afrique Française, various articles on the French occupation of the Tibesti and on Kufara in: *Renseignements coloniaux,* 1916, p. 173; 1917, p. 193; 1920, p. 69; 1921, pp. 6, 41.

DALLOUI, *Compte rendu d'un voyage au Tibesti,* T.I, partie géologique, in course of publication.

FORBES, ROSITA, "Across the Libyan Desert to Kufara," *Geog. Journal,* London, Vol. LVIII, 1921, pp. 81, 161.

GARDE, G., *Description géologique des régions du Tchad,* Paris, 1911.

HASSANEIN BEY, "Through Kufra to Darfour," *Geo. Journal,* Vol. LXIV, 1924, pp. 273-353.

NACHTIGALL, GUSTAV, DR., *Sahara und Sudan,* Berlin, 1879.

PELLEGRIN, J., "Poissons des Pays-Bas du Tchad," *Comptes Rendus de l'Académie des Sciences,* Jan. 19, 1920.

ROHLFS, G., *Kufra,* Leipzig, Brockhaus, 1881.

TILHO, JEAN, LT. COL., *Société de Géographie,* Vol. XXXVI, with map.

——*Comptes Rendus de l'Académie des Sciences,* CLXVIII, 984, 1081, 1169, 1236.

XIII
Fezzan

SEPARATING those two great volcanic massifs of the Sahara, the Tibesti and the Ahaggar, there is a deep depression which forms a wide defile, clear-cut and radical. It has a north-south orientation and lies almost on a meridian line with the large coastal indentation known as the Gulf of Sidra or the Syrtis Major. This extends between Misurata and Benghazi, and forms the deepest indentation of the Mediterranean coast. There are two large faults which inclose it on each side, and which we may designate as the Misurata and Benghazi faults. The westerly, or Misurata fault, is accompanied by eruptive occurrences in the vicinity of Sokna, and splits off the eastern extremity of Jebel es Soda, the famous Black Mountain which was the Mons Ater of the ancients.

According to Monsieur Bernet, a geologist who has given us the most recent study of the composition and structure of Tripolitania, these two faults, or two parallel systems of faults, furnish the explanation of the subsidence of the coast at this point; and he also estimates that they extend deep into the interior of the continent, in which he is undoubtedly right. For the whole center of the Sahara, between the Ahaggar and the Tibesti, has subsided along the line of this double system of faults, forming a definite separation between the two massifs.

On the floor of the depression we find certain irregularities which have an east-west orientation, roughly at right angles to the north-south strike of the faults. One of these is Haruj es Sod, which tends to link the Siwah cliff with the Black Mountain at the extremity of Tripolitan Jebel. Others are found farther south, at the point crossed by the caravan trails, where there are narrow sandstone plateaus known by the name of Tummo. These extend from northwest to southeast, and betray some kind of link between the Ahaggar and the Tibesti. But even Tummo hardly reaches an altitude of 2,300 feet, and in many places is pitted with enormous hollows, some of which are over 1,300 feet deep and less than 1,000 feet above sea level.

Not only do the topographical slopes of the two mighty flanking massifs incline toward these deep basins, but so also in a measure do the thick beds of the sandstone and limestone plateaus themselves. Consequently a considerable portion of the water reserves is conducted toward these low points, and the result is an abundance of water most surprising for the desert. It supplies many oases, scattered and in groups. To this whole strip of low-lying terrain, furrowed with deep hollows and strewn with oases, which cuts transversally across the Sahara from the Gulf of Sidra southward to the Sudan, we shall here give the general name of Fezzan, although it is the largest and most compact group of oases to which the term is more particularly applied.

The plentiful oases naturally mark out an ideal route for caravans. Considering also that they lie on a line with the southernmost point of the Gulf of Sidra, the place where the desert is narrowest and the coast approaches nearest to the Sudan, it is obvious that the trans-Saharan traveling distance is here shortened by 200 or 300 miles. Hence Fezzan offers not only the shortest but, by reason of the frequency of watering places, the easiest caravan route from the Mediterranean into Central Africa.

With the exception of the Nile route, Fezzan is also the oldest and most historical of the trans-Saharan communication lines. Its early importance is shown by the recurrence throughout history of certain geographical names connected with the region, and their persistence up to the present day in modified but easily recognizable form. The Black Mountain, as we have said, was the famous Mons Ater, and the ancient country of Phazania is plainly the Fezzan of today; while Jerma, which was formerly Garama, its original capital, preserves the illustrious name of the Garmantians.

The antiquity of Fezzan is easy to trace. The greater part of the Syrtes were under Carthaginian rule, and it is plain that Carthage did not lack interest in trans-Saharan commerce, for the Fezzan route was known to Herodotus five centuries before Christ. Succeeding Carthage, the Roman Empire established more or less of a domination over the province of Phazania itself and sent military expeditions into it on various occasions. Monuments and other archeological traces of Rome are found throughout the district, even at Jerma, or as it was then known, Garama, the ancient capital; while history records two Roman exploratory expeditions which pushed through Phazania as far as the country of the hippopotami.

There were two routes which led, and still lead, from

Tripolitania into Fezzan, of both of which Pliny gives us detailed accounts. The longer of the two, but the easier because it is strewn with watering places, goes by way of Sokna and Mons Ater. In 70 A.D. the Romans under Vespasian discovered another one farther to the west which shortened the traveling time by ten days but was much more difficult. This one traverses the solitudes of the Hammada Homra, and was the one followed by Barth from Tripoli to Murzuk. It was by way of Fezzan that he successfully carried out what was the first scientific trans-Saharan journey into the Central Sudan; and after him came Rohlfs, Duveyrier and Nachtigall, who were all cordially received in Fezzan and made sojourns there. This is all quite natural, for such a country, whose very purpose of existence for some two thousand years has been as a lane of travel, would be sure to offer the first facilities to European exploration. Fezzan is the most open part of the Sahara, in which respect it is in contrast with the Tibesti area.

Fezzan was occupied by the Italian troops around the end of 1929 and the beginning of 1930, and since that time the province has seen great activity on the part of the Italian scientists, especially the archeologists. A map of the region has also been published by the Cartographic Bureau of the Colonial Ministry at Rome, while today the whole country has been made accessible by excellent motor roads. Three or four years of occupation, however, are rather too brief a time to give us the entire picture, so that our knowledge of the region still lacks something in completeness and accuracy.

THE FEZZAN OASES

We have at least a clear enough idea of the very important group of oases which constitutes Fezzan proper.

It spreads out over a very extensive area, reaching northward through Sokna almost to Tripolitania, and extending far to the south in the direction of Tummo. Its general topography stands out quite clearly, being linked with the foothills of the Tuareg or western massifs; and it occupies the lowest portion of the whole region, just at the outlet of the great valleys of the Wadi Shiati and the Wadi-esh-Shergui which come down from the west.

As is usual in the lower reaches of the Quaternary valleys, there is here an enormous development of dunes. This is the Edeyen Erg, an almost exact counterpart of the Algerian ergs of the Igharghar and the Saura. It is a humid, habitable erg, the water being found in the form of lakes. These are not temporary lakes or shotts, but permanent lakes of living water. They are often quite deep and the water is generally salt or brackish, although in some cases it is fresh. The best known of these lakes is the Bahr-el-Dud, or "Lake of the Worms," and its name accords with its reputation, for it breeds a species of larvæ which hatch out into dipterous or two-winged insects (Artemia Oudneii), but which in their larval state constitute a source of food for the natives. A lake of the same nature, crater-shaped and very deep, is found in the Igharghar Erg and is obviously an emergence of the artesian layer; it is probable that the lakes of Fezzan have a similar origin.

Throughout the region the water in general appears to be just below the surface or standing in sumps or sink holes. Duveyrier mentions seeing some artesian wells and foggaras in Fezzan, but these were manifestly exceptions, as was also the rather monumental apparatus shown by him above a sump for raising the water, which

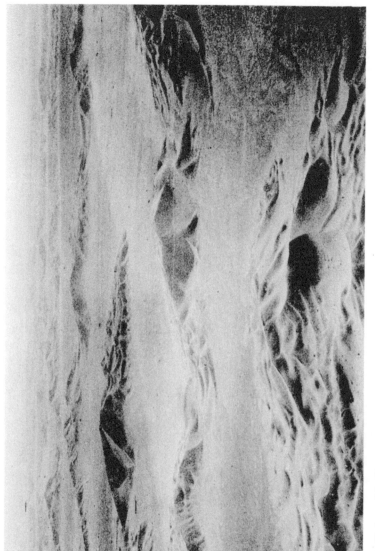

Air Photograph by Reygasse

THE GASSI TUIL IN THE GREAT ORIENTAL ERG

would represent real progress over the usual wretched lever wells that are in common use in the district and betray their ancient origin. It is evident that here in this land, where the water flows freely and spreads over the surface, the cultivator has had less need of a complicated irrigation technique, even though the means were known and at his disposal had he chosen to adopt them.

The lakes are nearly always encircled by palm trees, and Nachtigall praises the excellence of the dates grown in the region. Fezzan, for its abundance of surface water and the number of its palm trees, may perhaps have rivals around the rim of the Sahara; but there are no others in a corresponding situation, in the heart of the desert. In this respect it is unique.

The total population of the country was estimated by Barth, and later by Duveyrier and Nachtigall at some 50,000 souls—a very approximate figure. The Italian Captain Petragnani, who was a prisoner in Fezzan during the World War, was impressed by its decadence and misery, both due entirely to its present political state; and according to him the population must have been reduced to the approximate figure of 12,000. No more recent census seems to have been made by the Italian authorities.

By putting together the accounts of the ancient historians and those of the Arabian chroniclers, with the addition of native legends, Nachtigall has succeeded in reconstructing a history of Fezzan which on the whole is very satisfactory. It appears always to have had a distinct political entity, to have been a sort of little empire, although the embodiments of its history show successive influences from the north and from the south, and its capital has been changed from time to time ac-

cording to whether its masters were of Mediterranean
or of Sudanese origin. In the time of the Roman influence the Phazania of the Garmantians had its capital
at Jerma-Garama, and the dynasties of Berber or Arabian origin centered at Zuila. A Bornuan or Sudanese
dynasty has left deep traces at Traghen, where places
and streets still have names borrowed from the Kanuri
language, while Murzuk, the present capital, is of yesterday, being Turkish.

The resulting mixture of bloods is shown in the
present ethnological type, which is extremely confused.
The population of Fezzan comprises Arabs, Berbers,
Hausas, Tibbus and all the intermediate strains. The
predominance of black skins especially impressed Duveyrier, who on the strength of this alone built up an
elaborate theory regarding the Garmantians. According to him they were unquestionably pure blacks and
the propagators of a purely Nigritian civilization in the
Sahara. This theory would not be too far-fetched, if not
carried to an extreme and developed in too much detail.
Nachtigall seems to ally himself with it somewhat cautiously, since he claims to recognize in the present
average human type, as far as this can be distinguished,
a family relation to the Tibbu type. It must be remem-
bered, however, that a country situated like Fezzan has
doubtless never been able, even in the distant past of the
Garmantians, to avoid northern influences entirely; and
as we have already remarked, the most recent research
has shown us that the Garmantians in the beginning
were undoubtedly of white, Mediterranean, origin. At
the same time, Fezzan is essentially an oasis region; and
in the shadow of the palm trees, where malaria rages,
the white race never entirely succeeds in eliminating
the black.

Kawar and Bilma

There are other oases south of the Tummo Mountains which serve to put Fezzan into easy communication with Lake Chad, but these are of much humbler importance than Fezzan proper. The most noted of these are the Kawar oases, because of the salt mines, which are centered at Bilma. The salt is found here in a very pure state, and is carefully prepared by traditional methods; the commercial product is in compact cakes easy to transport. These salt works, situated on the finest caravan route of the Sahara, contribute much to its traffic; while undoubtedly they themselves would be less prosperous if they were located elsewhere. As we have already said, these Kawar oases, as well as Gatron and certain others of southern Fezzan, are definitely Tibbu as regards population.

Bibliography

Barth, H., *Travels and Discoveries in North and Central Africa,* London, 1852-53.

Bernet, E., "Contribution à l'étude géologique de la Tripolitaine," *Bulle. Soc. Géol. Fr.,* 1912, p. 385.

Duveyrier, Henri, *Exploration du Sahara,* Paris, 1864.

Nachtigall, Gustav, Dr., *Sahara und Sudan,* Berlin, 1879.

Petragnani, Enrico, "Quatre ans de captivité au Fezzan," *Renseignements coloniaux du Comité de l'Afrique Française,* April, 1922.

Rohlfs, G., *Quer durch Afrika,* Leipzig, Brockhaus, 1874-75.

Sahara italien, Le. Official guide book to the Italian section of the Saharan Exposition at the Trocadéro, May, 1934. Contains complete list of work done by the Italians, and several very curious photographs.

XIV
The Tuareg Sahara

HE whole remaining portion of the Sahara westward from Fezzan to the Atlantic Ocean may be considered as a single region, since there are certain general features which characterize it throughout. What chiefly distinguishes it from the Oriental Sahara is its much greater altitude, which if it were possible to calculate in actual figures would certainly prove to be much higher than the mean elevation of the eastern half. The general geological composition of both sections is practically the same, being made up of peneplanes of ancient rocks, often with sandstone or limestone plateaus superimposed. But the occidental half shows evidences of a recent upheaval; portions have been raised or tilted, others have subsided, and new conditions have been created throughout.

To be sure, the highest peak of the Sahara is Emi Kusi of the Tibesti range and is found on the eastern side; while the Tibesti itself, rising isolated and abrupt from the midst of the vast surrounding depressions of Fezzan, Borku and the Libyan Desert, is all the more arresting by reason of the sharp contrast which is formed. But it is, after all, the lowlands that make up by far the greatest percentage of the Oriental Sahara;

and the Tibesti in actual mass is not by any means comparable to the mighty Tuareg massif, which furthermore is only one of many mountainous masses dominating the Occidental Sahara. Moreover, the imposing Emi Kusi owes its impressive height to the fact that it is not only a volcano but one which is still fresh and intact; while the volcanoes on the western side are much older and are worn down and defaced. But taken as a whole, the occidental mountains predominate both in bulk and in numbers; and it is the massifs that form the outstanding feature of the western desert.

The principal of these massifs is the Ahaggar, whose summit is a kind of eroded platform with lava fields predominating. The platform itself is called the Atakor of the Ahaggar by the Tuaregs, or the Kudia by the Arabs; the two names are used interchangeably. It is 155 miles in greatest diameter, and maintains an altitude of some 6,600 feet in all parts, with extinct volcanoes jutting above the general level to almost 9,900 feet. The elevations surrounding the plateau are all high, diminishing progressively in slopes imperceptible to the eye. [See figure, p. 182.]

The Ahaggar is prolonged northward by other very extensive occidental massifs which include the Tassili, the Muidir and the Ahenet, all averaging some 3,300 feet in altitude; and still farther north the Tinghert, the Matmatas, the Tademaït and the Ugarta ranges rise in rugged prominences as high as 2,300 feet and form a chain which reaches to the Atlas. To the south a liaison is established between the Ahaggar and the Sudan through the massifs of Aïr and Adrar des Iforas, with altitudes of 5,600 and 3,300 feet respectively. Farther to the west there are still other massifs which from a distance dominate the Atlantic coast. Of these, the

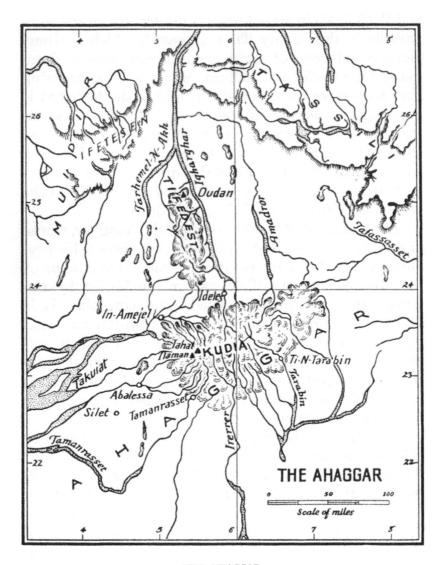

THE AHAGGAR

Eglabs have an altitude of some 2,300 feet, while the Adrar of Mauretania reaches almost to 1,650 feet.

Thus we see that the Occidental Sahara bristles everywhere with mighty massifs which either adjoin or continuate one another. It is also in this half of the desert that we find the Quaternary wadi systems still recognizable and almost coherent, covering great areas with their compact networks. The Occidental Sahara, in other words, shows a very much younger desert surface relief than does the oriental portion. As we have already said, this is a condition favorable to the diffusion of pasturages and consequently to nomadic life.

Moreover, this region being entirely dominated to the north by the Atlas range, the desert here is in open communication with the extensive northerly steppes, which are a great reservoir of the nomadic Mediterranean races. Thus, since the very structure of the Occidental Sahara, and the remarkable development of the old valleys with its resulting abundance of pasturages, offer facilities to this type of life, it is natural that the nomads of this territory should easily dominate the sedentary folk and hold them closely subjugated. This situation is exactly the reverse of that which we found in the Egyptian Desert; but it holds true throughout the western Sahara, where the nomads have not only extended to the very edges, but have even overrun the borders and penetrated into the Sudan.

The Occidental Sahara is a world unto itself, but it has always been more or less closely associated with the life of the Atlas. All through history the great nomadic tribes of Saharan Berbers and the Arabs of the Atlas, in spite of a bitter hatred for each other, have allied themselves against their common enemy, the sedentary people, and in this they have given each other mutual support. But while the part played by the Arabs is an important one, it is limited to the rim of the desert, for they range only along the foot of the Saharan Atlas and in Mauretania. The heart of the occidental desert belongs to the Berbers, and more particularly to that curious Berber tribe known as the Tuaregs. It would be fitting therefore to designate the whole occidental desert as the Tuareg Sahara, which would not only facilitate the exposition of the region but would further emphasize what is a truly remarkable phenomenon in human geography.

The Algerian Sahara

The Algerian Desert which occupies the eastern portion of this region, extending southward from Algeria and Tunisia to the elbow of the Niger, is by far the best-known part of the whole Sahara, because it has been under military occupation for the last thirty-five years. The physical aspects of this section have already been studied in considerable detail, except for its oases, which with the exception of those of the Egyptian Desert are not only perhaps the most interesting to be found anywhere in the Sahara, but have also been the most thoroughly investigated. The nomads here depend upon the support of these oases for their living, as is almost universally the case where there is a strong nomadic population.

Ghadames and Ghat

At the eastern extremity of the Algerian Sahara there is a small group of oases which includes Ghadames and Ghat, and which is of indefinite affiliation. Considering the fossil wadi systems, these two oases both

belong to the basin of the Igharghar, and in this connection they are undeniably related to the Algerian
Desert; but at the same time they are quite as undeniably related to Fezzan or the Tripolitan Sahara, since
they exactly mark the political boundary between the
two countries, and particularly as they lie on the Tripolitan side. In this group we have once more the am-
biguous situation of two oases located on the frontier
of two provinces, as was the case with the Kawar and
Gatron groups of southern Fezzan.

Ghadames, on the eastern edge of the great Igharghar Erg, lies in the bed of a wadi which comes down
from Jebel Nefussa and which certainly, before it was
buried beneath the dunes, once joined the Lower Igharghar. The geological conditions here have been studied
by Pervinquières, who demonstrates that the water of
Ghadames is artesian, which is the case with a number
of the oases bordering this erg, particularly those of
Jerid and the Wadi R'ir on the opposite side along
the Tunisian and Algerian borders. The difference is
that here the water is not well water, for Ghadames is
supplied by a fine natural spring which has always
been at man's disposal without seeking or effort.

This probably in some measure accounts for the historical antiquity of Ghadames. Under the name of
Cydamus, which is easily recognizable, it was partly
under Carthaginian, and later partly under Roman
sway. Duveyrier has given us a sketch of a bas-relief
that is obviously of Egyptian origin; while an inscription has been found in Greek characters and those of
some unknown language; and another in Latin mentions the Roman garrison, a detachment of the Third
Augustan Legion. Ruins have also been found of an
indeterminate nature, but which Duveyrier believes to

be analogous with the Roman ruins of Garama-Jerma
in Fezzan.

It would also seem that Ghadames was connected
with the early trans-Saharan commerce from the Mediterranean coast, and was likewise a base from which
the various influences were disseminated through the
desert. An evidence of this has been shown at a very
remote spot called Tabelbalet on the fringe of the erg
halfway between Ghadames and In Salah. Here a
number of cone-shaped stones have been found which
bear a rough configuration of a human face and recall
the idol stones of the Phœnicians, indicating a trace
of Carthaginian influence that must have reached it by
way of Ghadames. Tabelbalet would seem to have been
an outpost on one of the ancient routes through which
the Mediterranean commerce passed. The excavations
of 1934 at the Tomb of Tin Hinan have confirmed and
justified this idea.

At present the Ghadamesians maintain regular relations with Lake Chad and the Niger, and Duveyrier
much admired their enterprising spirit and their commercial organization. He was also astonished to find
that they used not only their own true Berber dialect
and Arabic, but also the Hausa tongue. Perhaps in
all this there might be traced some legacy from the
past.

Ghat, the second of the two oases which comprise
this group, lies much farther to the south. It is located
deep in the interior of the Sahara, on the same parallel of latitude as Fezzan and not very distant from it.
It has springs whose artesian character is attested by
the proximity of wells, as also by the fact that the
valley in which it lies is surrounded by the Tuareg
sandstone plateaus, a natural reservoir of the aqueous

layers. Ghat is at the very foot of their slopes and is only some 2,300 feet above sea level.

The orientation of the valley as well as that of the crests which skirt it is definitely north-south. A prolongation of the line between Ghat and Ghadames passes exactly through the Syrtis Minor and along the Tunisian coast of the Gulf of Gabes, marking also the eastern extremity of the Atlas range. It is quite natural to wonder whether this may not be another of the great sub-meridian accidents or faults along which the Tuareg massif has subsided toward Fezzan. Indeed the existence of such a fault at the oasis of Janet (Fort Charlet), which is only some 50 miles west of Ghat, has long since been recognized, first by Conrad Kiliau and afterwards by others. If this is so, may it not have some bearing upon the emergence of the artesian waters in the two oases?

Ghat does not seem to have had an ancient past comparable to that of Cydamus; it would probably have retained the memory of its founding only for some four or five centuries, and in any case its history is now lost in obscurity. At present Ghat maintains natural relations with Fezzan and is a sort of outpost for it, although until the recent and very strong Italian occupation it was actually under the domination of the Ajjer Tuaregs of Tassili. It lies on the direct route from Ghadames to the Sudan, one of those that was followed by Barth. This route, using the wells of the Wadi Tafassasset, connects Ghat with Aïr and so with the Niger.

THE ALGERIAN OASES

The Saharan oases seem to be divided into two general categories. There are those where the water is at the surface and available for man's purposes without

effort, such as we find in the extensive provinces of Fezzan, Borku and even Kufara, and of course in the Nile
Valley. And there are others where it has required extensive labor in the building of artesian wells or foggaras to exploit the waters which lie buried in the
ground. The oases of the Algerian Sahara belong in
the main to this latter category, for the rarity of springs
and easily accessible water is one of the principal characteristics of this region.

There are certain notable exceptions to this rule,
such as the oases of Taghit and Beni Abbes at the
western extremity of the Algerian Desert. These lie in
the Saura region not far from the Atlas Mountains, and
are supplied by fine springs of running water. Another
group, especially noteworthy in that it is something of
a curiosity, lies nearer at hand. These are the oases of
Suf, also called by the Arabic name El Wad, on the
northern border of the great Igharghar Erg near the
Tunisian frontier. The water here is found in an extensive layer at the surface of the ground just beneath
the sand, and each garden is a funnel dug in the sand
to the level of this layer. In this locality the gardener's
efforts are directed not toward irrigating his cultiva-
tions, but in keeping back the sand which threatens
them by caving in around the sides. [See illustration
facing this page.] Near Suf there are also the beautiful
Tunisian oases of Nefta and Tozeur, often visited by
tourists, which are supplied by fine natural springs.

But these are exceptional cases. In order to utilize the available water reserves in the great majority
of the oases of southern Algeria, considerable subterranean labor has been necessary. In this respect they
resemble the Egyptian oases of the Libyan Desert. In
fact these two regions, at opposite ends of the Sahara,

OASIS OF SUF, REGION OF EL WAD

Air Photograph

seem to be the only ones where this has been the case. And in spite of their wide separation there is such a marked similarity of techniques in use that the fact seems worthy of attention.

In this connection we might remark that throughout this region there is strong evidence of early Egyptian influence, such as we have already mentioned as having been introduced into the Sahara from Thebes by the oasis route through the gateway of the Isthmus of Siwah. The connection with Siwah, which was the center of the worship of Ammon, is particularly clear, since most of the traces indicate the influence of this cult. Corippus and the other ancient writers stressed the importance of the cult of the ram among the Saharan tribes, and Martin has published a photograph of a stone idol with a ram's head which was found at Tamentit in Tuat. Rock engravings have also been found at various points in the Saharan Atlas as far west as Figuig, with representations of a ram's head surmounted by the solar disc flanked by *uroeus,* which is obviously Ammon. Moreover the natives of the Wadi R'ir attribute the origin of their artesian wells to "Dhu-l-Karnain," which signifies "the Two-Horned," and is the name by which Alexander the Great is known in the Koran—but of course Alexander considered as the incarnation of Ammon, the god with the head of a ram. It is doubtless these early contacts with Egyptian civilization which explain the curious similarity of irrigation techniques and devices used in the two regions.

The Algerian oases have all been thoroughly investigated and are therefore very well known. In spite of certain characteristics which they all have in common, they are divided into two clear groups: those of the east, which are supplied by artesian wells, and those of the west, which are irrigated by means of foggaras. The choice is determined by the geological conditions, since for the maintenance of an important oasis it requires not only water, but water under certain conditions of output and economic exploitation. Unless there is sufficient pressure to cause the wells to flow, they will not suffice, even though they may be inexhaustible.

The eastern oases are more or less grouped together at the bottom of the depression which lies at the foot of the Atlas—Wargla, the most southerly of them, being not more than 185 miles from the foothills. In this region we have the immense and very regular syncline of the Lower Igharghar Valley, where all the beds are hull-shaped or spoon-shaped, from the Cretaceous, which forms the base, to the thick sedimentary layers that cover it. Natural springs are not entirely lacking here and take the form of small, often crater-shaped lakes which have an extraordinary depth for the tiny area of their surface. They may be as much as 100 to 130 feet deep, and constitute more or less obstructed vents of the deep water layer. They are stocked with fish of a residual tropical fauna, the most numerous variety being the catfish. The natives give these springs the name *bahr*, which literally signifies "the sea," although it is currently applied to all deep spring waters. These *bahar*, infrequent and brackish, are of no practical interest today, though perhaps they may originally have furnished the idea for the artesian wells which now with their gushing water supply all the finest oases, especially those of Jerid and Nefzaua in Tunisia and those of the Wadi R'ir and Wargla in Algeria.

The native methods of well making in this region, which have not yet entirely disappeared, or at any rate are still known, form an interesting contrast with the already well-advanced methods in use by the Egyptian well makers at the time of the advent of the Europeans. In place of the highly evolved tools of his more civilized confrere in Egypt, the well maker in the Algerian oases, outside of a pickax and *kuffin,* or bale, had only his bare hands and was forced to supplement the meagerness of his equipment with his own manual labor. The well makers here were more than a guild; they were a distinct tribe in which not only were the traditions handed down from father to son, but also a kind of atavistic predilection and even a physical adaptation for the trade. They were able to stay under water for an astonishing number of minutes and at the bottom of the well could endure the pressure of a column of water of extraordinary height. Moreover they accepted all the hazards of an appalling vocation with the calm stoicism of long familiarity. Evidently it was they, rather than their more advanced colleagues of the Libyan Desert, who kept intact the traditions of the early Egyptian well makers so much admired by Olympiodorus. [See illustration, p. 152.]

In Algeria the French occupation is now already well established, and the oases scattered at the foot of the Atlas are easy of access, especially since the railroad has been pushed as far as Tuggurt, capital of Wadi R'ir, and there is some question of extending it 125 miles farther south to Wargla. Needless to say, therefore, the wells of today are bored by the latest European methods, and under the palm trees irrigated by these artesian wells we find, after half a century of French organization, a population officially estimated at some 200,000 souls. They harvest and export what are probably the finest dates in the world, the famous "daglat nur," a fruit *de luxe* whose very existence presupposes a long selectivity based on antique traditions of gardening and consequently a very early civilization.

The distribution of the western oases is quite different from that of the eastern ones. Instead of being gathered in a group they are aligned in a single rank one after the other, forming a gigantic green ribbon which stretches across the Sahara. Starting at Figuig on the Moroccan frontier and ending at the oasis of In Salah, it makes an uninterrupted route from the Atlas to the foothills of the Ahaggar, in the very heart of the desert. It is divided into sectors which are, in order, from north to south: the oases of the Saura, of Gurara, Upper and Lower Tuat, and Tidekelt. This single thin strip of verdure is 745 miles long, and is called by the Arabian writers the "Street of Palms." They also, with a certain amount of oriental exaggeration, say it is so long "that a female caravan camel, mated at the beginning of a journey, will have time to bring forth her young before arriving at the other end."

Without a single exception, the oases of this system all have one characteristic in common; they consistently mark a geological boundary between the peneplane of ancient rocks on one side and the Cretaceous and Tertiary plateaus on the other. The line of oases follows every slightest deviation of the geological contact line, and the relationship between the two is obvious. With their beds sloping gently and regularly toward the oases, the great Cretaceous and Tertiary plateaus absorb the major portion of such rains as fall; and the extreme rarity of these at any given point is compensated by the immensity of the receptacle basins, which give back this water through leaks in their outer rims. These leaks, however, are not sufficient for irrigation

purposes unless they are cleaned out and primed;
and man, in order to exploit these sources, has had
to intervene, which he has here done by means of foggaras.

These are underground reservoirs, or supply galleries, which in their own way are as prodigious a labor
for the country as are the artesian wells. In some places
they attain a depth of 200 to 230 feet below the surface
at their heads; and they are spacious enough for a
small man to move about in them from end to end if
necessary. From place to place throughout their length
they are marked by ventilation shafts; and the outlets
of these, with their excrescences of heaped-up earth,
give the landscape an appearance of being strewn with
molehills. The total development of these galleries is
incalculable; for a single given oasis such as Tamentit
it may run to some 25 miles. All around any of these
oases the whole ground is undermined within a radius
of several miles, and one must tread cautiously in getting about. Such a system is comparable in importance
to the underground railway system of a great metro-
politan city. [See illustration, p. 234.]

These foggaras are of identical conception with
those of the Libyan Desert, of which the Geological Survey of Egypt has given us excellent descriptions. But
they are much more primitive in actual construction,
for at such places as Tuat we find nothing corresponding to the stone walls of fine masonry of Roman workmanship that are encountered in the vicinity of Baharia. The workmen in the occidental oases has had
little to work with except his own body and his bare
hands; he has supplemented his lack of tools with an
instinctive ingenuity and an animal-like tenacity. He
has become a human mole.

Thus as we go from the eastern to the western groups in Algeria, we find that it is not the human animal who changes, but the geological conditions. The human mole has simply known how to adapt himself to these conditions as he found them. But he has not arrived at this without some blundering, however. In certain of the western oases on the frontier of Tuat and Gurara we find native artesian wells which should give water in abundance but which fail to gush, since evidently, by reason of the structure of the subsoil, there is insufficient pressure. These wells are isolated and abandoned, because they cannot be utilized.

There are certain other places in this occidental group, especially in the bed of the Saura, where the surface waters are maintained by the regularity of the floods which come down from the Atlas. This surface water is found standing in sumps or sink holes which form natural fluctuating wells. The device for utilizing these waters is ingenious, and it too reflects an ancient civilization. It is here called *khottara,* although it is exactly the same as the Egyptian *shaduf,* which has been so often described and pictured. The pulley which we would use is replaced by a long pole weighted at one end by a large stone which serves to balance it; and a great leather bucket with a long handle, called *dillu,* is perfectly adapted to the apparatus. With this device it is easy enough to draw a bucket of water; but in order to irrigate a garden it would require an appalling number of dillus. The irrigating is done at night, to reduce the loss by evaporation; and when we reflect upon the labor represented in the life of a man who, from dusk to dawn for three hundred and sixty-five nights of the year, must ceaselessly go through the motions of drawing a dillu, it is plain to be seen that such a primitive arrangement would be impractical for irrigation on a large scale. [See frontispiece.]

Nevertheless, there is one important group of oases in the Algerian Sahara which would seem at first glance to depend entirely on such wells as these. This is the M'zab, which does not properly belong to either the eastern or the western groups, lying midway between the two in the midst of the dreadful solitudes of the Cretaceous plateau. Its palm groves are planted in the deepest of the Quaternary wadi beds, to get as close as possible to the subterranean water; and the wells, although dug in the hardest of limestone, are nevertheless nearly 200 feet deep. This great depth makes useless even the true khottara by hand manipulation, and it is necessary here to employ donkeys or camels to draw the dillus. But the upkeep of these animals is costly. The M'zab, in fact, with its 43,000 inhabitants, can hardly be considered a self-supporting oasis. It continues to exist because the M'zabites are a different sort of people from the other oasis dwellers.

All the adult male portion of the population lives outside the desert, in the great cities of Algeria. They are merchants, money lenders and bankers, and accumulate vast fortunes. Like the Jews and Armenians, to whom they correspond, they are held together by the bond of an ancient religion; for they form a very exclusive Moslem sect, and are exceedingly jealous of their faith and their autonomy. For them the M'zab is a pleasure garden and a citadel, a whim and a costly necessity; a place to return to, but certainly not a place to remain. It would soon vanish into thin air if the financial prosperity of the tribe were to collapse.

For an oasis, in order to be self-supporting, must have running water at a level higher than that of the garden, which will flow down the slopes by itself and irrigate the foot of each palm tree without any effort on the part of man. This condition is fulfilled only by the artesian wells and foggaras, which, in short, represent an enormous conversion of capital into public works for the purpose of reducing manual labor to a minimum. The whole system constitutes a delicate masterpiece of economic and financial management to balance the net cost with the yield, and has required a meticulous calculation of economic factors in which not even the most trifling details could be ignored.

The dog, for example, is quite unknown in the oasis —not because he could not live there, for he has lived there in the past. In fact, according to Pliny and el Bekri, dog meat used to be a source of food in the oases of southern Tunisia. But to the Moslem the dog is an impure animal, unfit for food. Evidently he disappeared at the same time that the introduction of Mohammedanism rendered him useless for slaughter, when he was eliminated as a worthless mouth to feed. Another example of the careful attention to small details is the large number of very well-tended latrines to be found in the oases, for the reason that the excrement is much too precious to be wasted.

The ownership and use of the water are also determined by a meticulous and ingenious traditional legislation which presupposes a long elaboration throughout centuries or even thousands of years. This would form the basis of a whole study in itself, and is one that has just been touched on by Brunhes. Nor must we neglect to mention some of the cleverly conceived devices used to measure the water drop by drop and minute by minute in apportioning it among its common proprietors. One of these is a kind of water clock and measures the time; another is in the form of a comb fixed like a goose claw at the intersections of the smaller irrigation canals, dividing the total volume of water through its teeth and acting as a gauge calculated on the pro-rata rights. All of this, of course, including customs and instruments, has had its origin in the ancient oriental civilizations. [See illustration, p. 88.]

The architectural aspect of the small villages under the palm groves is also noteworthy. The settlements are *ksars,* or fortified towns, for the Sahara is not a place where one may sleep with open doors. With very few exceptions both walls and dwellings are built of clay or crude bricks of hardened mud. The wretchedness of the materials makes the complexity of the constructions seem all the more remarkable, for the buildings are several stories high and have inclosed stairways, while the streets are roofed-over arcades. The whole has an urban aspect, and the social life is urban as well; there are markets, shops, promenades, cafés and recreation places. All of which is indispensable to the nomad, who demands of the oasis what the sailor demands of a port of call: easy reprovisioning and the gross slaking of long abstinences. A ksar, however tiny it may be, is never a village, but a city in clay. The Babylonia of Herodotus was built upon this very model. [See illustration, p. 198.]

Whichever way we turn, we are here in the midst of the early millennial civilization of the Orient; and this is all the more remarkable because today it is entirely in the custody of a few poor Negroid savages. The majority of the natives of the oases are *haratin,* a word which seems etymologically to signify "cultivators" or "peasants," but which in current usage is applied exclusively to the Negro cultivators. This association of ideas is quite natural, for under the shade of the palm trees the vitality of the white race is sapped by malaria which prevents them from making any physical effort, while in crossbreeding the white blood tends to be eliminated. Thus we find that the "Ksurians," the inhabitants of the ksars, are in the majority Negroid.

This must not, however, be taken as a substantiation of the theory that the whole Sahara was formerly the domain of the black race. The ksurians are not related to the Tibbus and give no evidence of being aborigines. Not only have they as a whole no common and ancient tradition, but individually each seems to have the memory of a grandfather or an ancestor who came as a slave from some place in the Sudan. They speak only Arabic and Berber, although in certain places a kind of Sudanese dialect is used. But this is distinctly a dialect, a mingling of some twenty different Negro tongues. All of which would seem to indicate that the haratin of the occidental oases, comprising some fifty thousand souls, are not an indigenous population but a residue left from centuries of an uninterrupted importation of black slaves. If there is a still more ancient substratum, it is no longer discernible.

This is not very surprising, for we are here dealing with that portion of the Sahara where the white nomads, supported by the Maghribi of the Atlas, swept everything before them. Also, as we have already pointed out, the oases of the Algerian Sahara are of authentically recent foundation, dating in the case of Gurara from about the sixth century A.D., while certain of those in the vicinity of Tidekelt are very much later, dating only from the eighteenth century. Consequently, in this region where the cultivation of the palm trees in

Air Photograph

TYPICAL SAHARAN KSAR: TEMACIN IN THE WADI R'IR

the majority of cases is closely associated with skillful irrigation of a definitely oriental origin, the most natural supposition is that it was the great invasion of nomadic camel herders that was responsible for the present system of intensive agriculture.

THE NOMADS

These nomads, who in all probability were the founders of the oases, are in any case the present masters of them. As a general rule they have an individual ownership of the palm trees and appear at the moment when the dates ripen to conduct or oversee the harvesting personally. The haratin are no more than leaseholders of a small percentage of the harvest, *khammis* as they are called in Arabic from a word that signifies "five," because they have a right to a fifth part of what they produce. In addition to this, each group of oases is under the political domination of one sovereign nomadic tribe, on whom it depends for military protection against other warlike tribes. The fortifications of the ksars are for protection only against raids and attacks from neighboring ksars, and in no way indicate any idea of resistance to the sovereign tribe. The whole organization much resembles that of Europe in the Middle Ages, when the serfs depended upon the protection of their barons.

This servility of the ksurians appears natural enough when we consider just what sort of men the nomads are. In the first place they represent the only armed force; strife is their trade and their daily bread. Furthermore they are of another race, being all incontestably whites. To be sure, it is not actually impossible for the black race to become acclimated to the great desert spaces beyond the oases and a dry climate of

violent extremes, as is demonstrated by the case of the
Tibbus. But the Mediterranean white is already at
home in the desert and his organism is perfectly ad-
justed to its rigors.

The very nature of their life has made these Saharan nomads all picked men. The occidental desert, as
we have described it, is furrowed with routes which are
extremely difficult to travel but practicable under
necessity; and it is the hardships of these routes that
have made the nomad what he is. We should guess this,
if only by a comparison of the mehari saddle used in
the Occidental Sahara with the Sudanese saddle of the
Egyptians. The latter is very large, fitting over the
hump and covering the entire back of the animal; it is
comfortable and almost permits one to recline in it, but
it is very heavy. The saddle used in the western desert
is the *rahla,* a word which means literally "the travel-
ette," and is nothing but an assemblage of four planks
set in front of the hump. It has no stirrups and the
rider's seat can be retained only by placing his bare
feet on the camel's neck. The thing is a marvel of light-
ness and suited to the appalling demands made upon
the beasts under circumstances where even a few
pounds more or less are of tremendous importance. But
only men in the finest physical condition could adapt
themselves to such primitive equipment. The constant
circuits of the desert, which impose upon the human
organism an extreme physical effort and at the same
time demand an extreme sobriety, have given them
magnificent bodies, slender and finely muscled. A typical example of the Saharan nomad is the Massinissa
of the Latin writers, who at the age of eighty is said
to have conducted a cavalry charge and begotten a
child.

Their mental faculties correspondingly reflect their environment. Let us consider what these Saharan routes are like, where death by thirst must be the inevitable outcome of a single instant of inattention, a momentary weakness or lack of presence of mind. Nor is thirst the only danger; there is also man. For this is *balad-al-khuf,* the country of fear; or *balad-as-sif,* the country of the sword. An unknown track crossing the path may perhaps give warning of an ambush; and it is unsafe to linger at the well-known watering places where one may be spied upon. The water bottle is hastily filled; but camp is made much farther on, after a detour calculated to mislead the impending and always probable pursuit.

Such a life results in an acuteness of the eye and mind that is an amazement to Europeans. An absolutely illiterate nomad, questioned by an explorer, will draw an intelligible map with his finger in the sand; he has a highly developed sense of topography and direction, for these are matters of life or death to him. He will also recognize, by the imprint of a naked foot, a certain member of a certain tribe, as infallibly as a European police agent identifies a criminal by his finger prints. As for his character, there is little need to say how deeply it is tinged by the always present shadow of violent death.

Such is the individual. But we must also consider him in relation to his tribe, and the bonds which unite him with the other members of it. They are exactly equivalent to those which bind our soldiers together under military discipline, for a nomadic tribe is a regiment born. Against such men as these it is natural enough that the Negroid folk of the oases should never even conceive the idea of resistance.

Chaambas and Tuaregs

There are two distinct groups of nomads in the Algerian Sahara, the Arabs and the Berbers. The two have nothing at all in common except the same adaptation to the same type of life. Apart from this they differ in every respect, including language, customs, weapons and garb. An even more serious breach is affected by their religion, for they represent different sects or degrees of the Moslem faith and consequently are imbued with imperishable hatreds for each other.

The two groups live apart, in separate regions of the desert. The Arabs hold the northern section along the foot of the Atlas mountains and in close communication with the Arabized Maghrib; the most powerful tribe is that of the Chaambas, who rule over the eastern group of oases, particularly Wargla. Their pastures are in the wadis of Tademaït, but they are especially at home in the pasturages of the erg, and their camels are said to be so accustomed to the sand that their feet are more tender than those of other varieties and more easily wounded in the stony desert.

These Chaambas, except for their Saharan characteristics, are not distinctly different from other Moslems of the Arabic tongue. Inhabiting occupied territory and thus in long contact with the French, they have to a great extent furnished the personnel for the meharist or camel corps; in fact it might almost be said that the French mounted troops of the Algerian Sahara are really the Chaamba tribe mobilized. For they have furnished more than soldiers; they have introduced their own methods and brought with them the Saharan spirit. Under the guidance of their French officers these troops during the last thirty years have pacified the whole Algerian Sahara; but the Chaambas by themselves, before they were thus organized and directed, were never able to accomplish this. For centuries they remained encamped at the foot of the Atlas, abandoning the heart of the desert to their secular enemies, the Berber Tuaregs. [See illustration, p. 130.]

The Tuaregs, unlike the Chaambas, have a marked individuality of their own. Representing only a small fraction of the human race, they have nevertheless a world-wide notoriety which may be partly owing to chance and the mysterious glamour of their desert background, but is certainly in a measure due to the remarkable aggressiveness which characterizes them and the exceptional originality that distinguishes them as an unusual human type.

In spite of the fact that they are Berbers and belong to the white race, they have certain traits in common with the Tibbus, in which connection we should recall that the ancestors of the Tuaregs—who, as we have shown, may have been the ancient Garmantians—wrested the country they occupy from a Negroid people who were probably cousins of the Tibbus. In any case, they now dress like the Tibbus, in Sudanese cotton stuffs of black or dark blue, and, what is even more striking, like them they also wear the *litham* or famous Saharan veil which masks the entire face except for the eyes and is never removed. Nothing in fact so much resembles the Tuareg silhouette as the pictures of Tibbus in the engraved illustrations of Nachtigall's book.

The Tuaregs, isolated for centuries in a lost corner of the world, are still closely linked to primitive humanity in an astonishing degree. They still make arm bands of polished stone and their ax hafts are of Neolithic construction. The litham has no hygienic purpose among them, but is a pure survival of animism, being worn not as a veil to protect the respiratory tracts against the desert wind, but to keep "bad spirits" out of the nose and mouth, which are the gateways of the breath and thus of the soul. Moreover we have learned from the 1934 discoveries in Tassili that representations of the human face were taboo among the Garmantians from whom possibly they are descended, and the veiling of the face itself might be part of the same taboo. The Tuaregs also have taboos which smack of totemism, and will not eat the great lizard, *uran,* which is such a favored article of diet in Algeria, because "it is their maternal uncle." They retain obvious traces of matriarchal rule; the male chief of a family is not the father but the maternal uncle, and succession is entirely in the maternal line.

Their language of course is Berber, but, what is even more remarkable, they are the only ones in the world to write it. Among them and nowhere else the ancient Libyan alphabet, called *tifinar,* is still in use. They still habitually carry the arm dagger, just as Corippus described it; they constitute, in fact, a last specimen of the Libyan, as if preserved under glass. Yet in apparent contradiction to all this, these primitive people in many ways are much more like ourselves than are the Arabs. They have more frankness and more curiosity, and are much easier to get on with. This is because they are less under the influence of Islam. Arabic, the sacred language of the Koran, is entirely unknown to them, and they do not keep the Feast of Ramadan, while their women have a freedom that is far more like our own feminism than like the Mohammedan customs.

Of all their ancient Berber heritage, the one thing

they have perhaps most faithfully guarded is an undying hatred of the Arab invader. This is a warfare that has never known a truce. It has been, however, an unequal warfare; for the Arab in his proximity to the Mediterranean has always been able to follow more or less closely the modern developments in armament. The Tuareg, on the other hand, without access to the outside world, had only his primitive traditional weapons. The coming of the French troops at the beginning of the twentieth century found the Tuaregs still equipped with the lance, the great shield of antelope hide, and the huge right-handed sword; the latter, having no point and used exclusively for striking or cutting down, recalls the glaive of the Gauls described by Livy.

This surely was an admirable outfit for warfare! Yet with only such arms as these, the Tuaregs for centuries blocked the Arab invasion and held the exclusive domination of the desert routes. Nachtigall recounts a battle in which the Tuaregs defeated the redoubtable Arab tribe of the Uled Sliman, wiping them out almost to the last man. It was in one of their characteristic surprise attacks, taking place just before sunrise; a close, hand-to-hand combat of incredible ferocity. The French column under Colonel Bonnier was annihilated in exactly the same way near Timbuktu.

This militancy of the Tuaregs is all the more remarkable because in actual numbers they are so few. Even the most powerful Ahaggar tribe cannot include more than three or four hundred meharists at most; and this is the tribe that is the most feared, which to be sure does not necessarily mean it is the largest. As for the whole Tuareg population, we cannot even attempt to estimate the total number, but it is certainly almost infinitesimal. Also we must remember that the Tuareg tribes, as is usual among the nomads, are dis-united and divided against themselves by eternal ven-dettas. The Ahaggars, for instance, who graze their camels in the Atakor, the Muidir and the Ahenet, and used to control the oases of Tidekelt, never get on with the Ajjers who graze in Tassili and are masters of the oasis of Ghat. Yet in spite of this lack of numbers and solidarity, the Tuaregs would undoubtedly still be masters of the Occidental Sahara had not Europe forc-ibly intervened. [See illustration facing this page.]

THE SUDANESE BORDER

In order to comprehend fully the force and extent of the white nomadic invasion of the Sahara, we must follow the Tuaregs all the way to the Sudanese border, where they were already intrenched when the Euro-peans arrived. This sector of the Sahara includes Aïr, Adrar des Iforas and the elbow of the Niger, all of which are really important outposts of the Sudan. No-where in this extensive borderland do we find any im-portant oases, and none at all of the type that flourish in the north under the Negro cultivators. The reason for this may be purely historical; or it may be that the climate, drawing so near to the equatorial zone, is al-ready too wet for the date to ripen. In any case, skilled irrigation and intensive agriculture, disseminated through the northern Sahara from Europe, never reached the southern portion of the desert—a fatal circumstance. Today only pasture grounds occupy what should be cultivated fields, and the Negro agricul-turists are pushed away by the nomads.

The most Saharan of the three provinces is Adrar des Iforas, at least in so far as regards population,

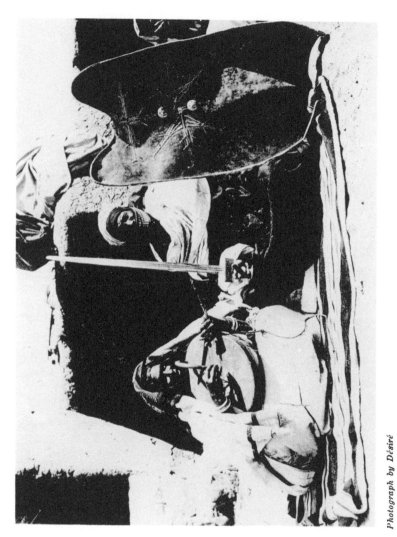

Photograph by Désiré

TUAREGS, MAN AND WIFE

which is exclusively Tuareg. The Iforas are a very ancient Berber people, whose name in the easily recognizable form of "Iforaces" is mentioned by Corippus; and they still retain among their legends a somewhat hazy version of the exploits of Kusaila, the old Aurasian hero who in 683 A.D. killed Sidi Oqba, the first Arabian conqueror. At present they speak a Tuareg dialect and recognize the supremacy of the Ahaggars, with whom they share their pastures in bad years. From the viewpoint of human settlement their Adrar is simply a prolongation of the Ahaggar; it is also the great Tuareg gateway for communication with the Niger, since farther to the west such communications are rendered most precarious by the widest extent of the Tanezrouft. Essentially the Adrar is a place of pasturages for the nomads, and such few insignificant palm groves as there are do not constitute important oases. In the way of vegetation and climate it is less Saharan than Sudanese, beginning already to show influences of the neighboring steppe.

Aïr is much more important than Adrar, for it forms a kind of crossroads of the great trans-Saharan trade routes. The route from the Ahaggar and the ones from Ghadames and Fezzan by way of Ghat all pass through Aïr. It has a certain number of straggling market towns, all of which are essentially commercial centers, beginning with Iferuan at the north and reaching to Agades in the south. Barth and Foureau both made long sojourns in Aïr. The number of its inhabitants must be nearly 20,000; the population is very mixed, but its base is manifestly Hausa and the Hausa language is understood by everyone. An important proportion of this population is made up of black Tuaregs, of mixed Hausa and Tuareg blood, but who naturally claim the Berber ascendancy as being the more honorable. There are also some white Tuaregs among them; and these by reason of their aggressiveness and prestige are the real masters of the province, in spite of the presence of a sultan at Agades. Or to be more exact, they were the masters up to the time of the French occupation.

The region around the elbow of the Niger is altogether different in every respect. During the Middle Ages this was the seat of some of the great Negro empires, one of which had its center precisely at the bend of the river. Its capital was Gao, whose ruins are found at the eastern extremity of the elbow where it is joined by the great valley of the Tilemsi, now practically dry but leading straight down from the Adrar des Iforas. This was the empire of the Sonroï, who are still found here meagerly populating the river flats as far as Timbuktu, but much degenerated. The present masters of the bend are the Tuaregs, or at least they remained so until the French occupation.

The Tuaregs found along the Niger are not the Ahaggars, but other tribes including the Aulimmidans and the Kel-Geress. They differ from the real Saharan nomads in many ways; for one thing, they are more numerous, because they lead an easier life. They have also had to renounce at least the exclusive use of camels, on account of the terrible epidemics among the camel herds caused by that host of microbes, the tsetse fly, which for a part of the year swarms on the bank of the river. For this reason they have here taken to the horse. But they are Tuaregs nevertheless; they retain the traditional costume, the language and the national spirit.

Along the Sudanese borderland the Tuaregs are in

close contact with their hereditary enemies, the Arabs of Mauretania. One tribe of these, the Kuntas, who it is true are more interested in religious than in military matters, are pushing north of the Niger and have already reached the slopes of the Adrar des Iforas. But it is the Tuaregs alone who hold the Niger on both banks; and who, moreover, in true nomadic fashion, are allowing it to go to waste. It is true that in the midst of the desert proper the nomads are forced to depend upon the oases, for they could not live by the desert resources alone; hence they encourage cultivation and protect the agriculturists. But in the steppe regions, where they can subsist without additional resources, they have no interest in sedentary pursuits and are sworn enemies of law and order, so indispensable to the peasant. This holds true in the north as well as in the south. Thus on the vast river flats of the Niger, where millions of men might well subsist, we find only a few herds of cattle tended by Sonroï herdsmen who live in an abject and ludicrous terror of their masters. When the French administration wanted to distribute rifles among these Sonroï Negroes, they replied, pointing to their agile legs, "These are our guns, if we have need."

The chief center of this region is Timbuktu, which for many years had a fabulous renown. Its name, in fact, echoed from one end of the desert to the other— as long as the town that bears it was known only by hearsay. Today we know that it is of but trifling importance in itself, and that even in its heyday it had only some 12,000 inhabitants, which now have dwindled to not more than 4,000. The basis of its prosperity has been the exploitation of the salt mines of Taudeni; and even now the great periodical event at Timbuktu is the arrival of the Azalai, or grand caravan, from the salt works. Taudeni lies some 375 miles to the north in the midst of the Sahara, and the salt industry is an enterprise that is entirely artificial. For Taudeni is not fit for habitation and the Negro workmen, who are imported and held by force, are killed off within a few years by its brackish waters. It is hardly possible that even under the most merciless régimes there has ever been an industrial hell comparable to this one anywhere on the face of the earth. Such an industry certainly does not form any real base on which to build a future, presenting no opportunities for either development or duration. Of course it is true that Timbuktu is also essentially an outpost for Jenne, the great commercial metropolis of the Niger, which is located much farther upstream in the midst of the Sudan.

In spite of existing conditions, there is reason to expect that this region around the elbow of the Niger will have the finest future of any portion of the Sahara, and one out of all proportion to its present wretched state. This future lies naturally in the river itself, a mighty watercourse flowing through the midst of the desert and bringing to it tremendous annual floods which overflow and spread across the terrain. It is truly a second Nile, lacking only management to cause it to fertilize a second Egypt. There is not another spot in the whole Sahara where such financial possibilities are indicated. With this in view, Timbuktu seems destined to become in reality what it used to be in the mirage of its remoteness, the great metropolis of the Sahara.

MAURETANIA

The extreme occidental portion of the Sahara, westward from the Saura and the Niger, is an immense country of which there is very little for us to say. A considerable part of the coast is Spanish territory and still unexplored; while the interior is Moroccan not Algerian territory, and the French occupation of Morocco has hardly more than succeeded in crossing the Atlas. It is not yet really established in the Sahara. The Meharist Companies of French Occidental Africa have made a commendable effort to open up the territory to the south, and Malavoy, Director of the Geological Service of Dakar, has made one interesting expedition into the region. But the area covered is still fairly limited and the necessarily fragmentary results have not yet been systematically worked up, although an excellent geological paper has recently been published by Metchnikoff.

Thanks to all this, and to the early explorations of Lenz and de Foucauld, we can at least begin to get some little idea of the general framework of the region. The northern portion is dominated by the massif of the Eglabs, an accentuated upthrust of the peneplane girdled about by great elongated ergs. We have slightly better information about the southern portion bordering on the Sudan, for we know the Mauretanian Adrar to be a plateau of red Devonian or Silurian sandstone, comparable to the Tuareg plateaus of Tassili, Muidir and Ahenet. From the viewpoint of human life, however, the part which is most interesting is precisely that which is most unknown.

We get only the merest glimpse of the great oases that lie along the southern edge of the Moroccan Atlas. The meharists of the Saura, in opening up their territory, have found their activities west of that river very much hampered by a rather unusual circumstance, for the Moroccan Berbers, who prior to the French occupation were masters of the palm groves of Tafilalelt and the Saura, abandoned them only after many bloody battles, and have still continued to constitute an annoying menace on all the routes west of the river. The meharists however, in policing the district from newly created posts in the Iguidi Erg, have greatly enriched our knowledge of the region. The important oases of Tafilalelt and Dra'a have now also been occupied by Moroccan troops, but this has been of hardly more than a few months' duration; and while we already have maps compiled with the aid of air photographs and published by the Cartographic Service of Rabat, we have not as yet any detailed study of the inhabitants.

In fact, without the efforts of de Foucauld the greatest portion of the Sahara that lies at the foot of the Atlas Mountains would still be blank upon the maps. Tafilalelt would seem to be a sort of tiny world, whose ancient capital of Sigilmessa played an important part in the Berber Middle Ages; but we have little data on these oases except for the notes of explorers like Rohlfs, de Foucauld and Harris, who passed through rapidly and often in hiding. As for the oases of Dra'a, de Foucauld considered them to be the finest in all the Occidental Sahara, but other than this we know very little about them.

We do know, however, that throughout all these oases, particularly in Dra'a, a large part of the population is made up of the haratin class. These haratin are a Negroid people with what seems to be a Berber dialect; but they seem also to have been established here from very early times. It has yet to be determined whether they are descendants of imported slaves like the Negro cultivators of the Algerian oases, or represent a truly aboriginal population such as we find in the Tibesti. Are these peasants of Dra'a perhaps the Melano-Getules of antiquity, a last remainder of the Saharan Negro? It is still an open question.

Some of the most important problems awaiting solution are those of the Atlantic coast. A study of the physical geography particularly might prove useful in throwing some light on a subject of world-wide interest: the question of the lost continent of Atlantis. Such theories as we have at present are based on the text of Plato and are very vague, although geologists and zoölogists now admit that there has been some recent continental subsidence in the ocean depths. But a detailed investigation of this portion of the coast might perhaps permit some interesting determinations in the matter.

Other questions still unanswered are those concerning the human race. It would seem that the Berber invasion must have reached the Sudan much more quickly along the Atlantic coast than it did elsewhere in the Sahara, for as early as Ptolemy we find mention of the Berber Sanhajas, or Zenagas, at least in the north; and it was certainly this great, well-known tribe that gave its name to Senegal. These Sanhajas are known to have been nomads and wearers of the veil, closely related to the present Tuaregs; they were also none other than the great Almoravides, who founded a mighty empire and conquered both Morocco and Spain. They seem, moreover, to have been the only exclusively Saharan tribe who ever played such a part in the history of the Maghrib; there is no parallel example. Yet we cannot readily see what geographical conditions made possible such a result. In any case, the Sanhajas now, together with their language, have almost completely vanished from the region which was their country of origin.

This fact should not particularly surprise us, for it

is general throughout the Maghrib. The same thing happened to the Berber Ketamas, who founded the Fatimid Empire in Algeria. Any Berber tribe who founds an empire is always in the long run consistently defeated by its own triumph; it becomes Arabized and gradually disappears. We know that sometime around the fifteenth century, after the overthrow of the Moslem power in Spain, the Maghrib was overrun by certain marabouts of the Seguit-el-Hamra, or Rio de Oro, who played an important rôle throughout the region, as missionaries of the Moslem faith and propagators of the Arabic tongue. About these people, however, we know scarcely anything at all.

Nor do we know very much more about the present inhabitants; except that today, in this country where the Almoravides originated, we find the Moors, for which reason it is now known by the name of Mauretania. The Moorish tribes not only speak the Arabic tongue, but are much more cultured and literary than the other Arab tribes; and their Moslem piety is far more strict. As we have remarked before, these two qualities seem always to be closely allied in Islam. Nor does all their devoutness and refinement prevent them from being bandits into the bargain; for the Arab tribes of the Atlantic coast, particularly the Uled Delim and the Regibat, are among the most dreaded of robbers.

This is all the knowledge that we have. What may have been the relations between the Sanhajas, or Almoravides, and the marabouts who followed them, or the Moors of today, is utterly obscure. In fact there is no corner in the whole Sahara, even in the Libyan Desert, that is more unknown than is the Rio de Oro to the present day.

BIBLIOGRAPHY

Besides several works already cited:

BRUNHES, JEAN, "L'Irrigation," Doctorate thesis.

Cartes de la Mission du Transafricain, Société de Géographie de Paris.

Comité de l'Afrique Française, articles on Mauretania, *Renseignements coloniaux,* 1912, p. 20; 1915, pp. 73, 118, 136.

CORTIER, M., *D'une Rive à l'autre du Sahara,* Paris, 1908.

—— *Mission Cortier,* Paris, 1914. On the Adrar des Iforas.

Croquis de l'Afrique du Nord, 1 : 5,000,000. Service Géographique de l'Armée, 1922.

FOUCAULD, C.-E. DE, *Reconnaissance du Maroc,* Paris, 1884.

GAUTIER, E.-F., *La Conquête du Sahara,* Paris: Colin, 1919.

GRUVEL, A., and R. CHUDEAU, *A Travers la Maurétanie occidentale,* Paris, 1909.

LENZ, O., *De Tombouctou au Maroc,* Paris, 1884.

MARTIN, A. G. P., *Les Oasis saharienne,* Paris, Challamel, 1908.

MENCHIKOFF, "Recherches géologiques et morphologiques dans le nord du Sahara Occidental," *Revue de géographie physique et de géologie dynamique,* Vol. III, Part 2, 1930.

MEUNIER, A., *Map of Central Sahara,* Service Géographique du Ministère des Colonies, 1917.

PERVINQUIÈRES, L., *Ghadamès,* Paris, 1912.

RESSOT, CAPT. A., "Considerations sur la structure du Sahara," *La Géographie,* I, 1926, 26.

ROD, RENNEL, *The People of the Veil,* London, Macmillan, 1927.

ROHLFS, G., *Mein erster Aufenthalt in Marocco,* Bremen, 1871.

ROLLAND, G., "Rapport géologique," *Documents de la Mission,* Choisy, 1890.

Territoires du Sud de l'Algérie, Les (official publication), Algiers, 1922.

PART V
CONCLUSION

XV
The New Sahara

N spite of the many gaps in our knowledge, the present picture of the Sahara in the aggregate is clear and coherent. The reason this is so is that North Africa in the last three-quarters of a century has become European territory, a fact of tremendous importance, and one whose consequences will reach far into the future.

The most radical transformation to date is that the Sahara has ceased to be an obstacle, a bulkhead interposed between Europe and Black Africa. This bulkhead has never been perfectly air-tight; it has always permitted a certain amount of traffic to pass through, but it was very little and extremely slow—depending exclusively on camel caravans, whose average speed was some two and a half miles per hour, and which in the course of a journey involving twelve to eighteen hundred miles required long rests in the oases and pasturages. Modern methods of transportation, which have so greatly shortened distances all over the world, have thus had an exceptional importance in the Sahara.

The World War, which launched in North Africa the Turkish and Senusi drives, more than anything else gave impetus to experimentation and the development of new methods of transportation and communication in the desert. The necessities of the defense against the Turks in Egypt and the Senusis allied with them in the French and Italian Sahara required the immediate use of inventions which hitherto had not been thought of practical value in the desert; but many of these experiments, having proved their usefulness and practicability, have been permanently adopted.

The most immediate need was for telegraphic communications, which problem was quickly and easily solved by the wireless telegraph. Stations were installed throughout the English, Italian and French Sahara, and at once gave such excellent results that they became the essential cogs in the defense mechanism. These emergency stations have become a permanent system, furnishing a powerful auxiliary for keeping peace, order and security in the desert by permitting every bit of news in the interior to be transmitted immediately to the local and central authorities, who can communicate and execute their decisions with the same dispatch. In this way it forms a kind of "nerve system" for the desert; but it has another distinctive value in that it permits the isolated European in the far-flung outposts to be in constant touch with the outside world, receiving daily news bulletins from the Havas or Reuter services. It thus acts as a powerful psychological aid in mitigating what in French slang is called "cafard," the neurasthenia of solitude. It has also brought in its wake a great need: the imperative demand for a dependable source of industrial power, never before felt in the Sahara. This demand must lead to interesting experimentation, possibly along the lines of utilizing in some way the desert wind.

The transportation of men and merchandise is naturally more difficult of realization, but great progress

has already been made along these lines as well. With
all the financial prodigality which characterized the
World War, the use of airplanes and automobiles was
put to the test for Saharan transport; and the results
with motor vehicles at least were far from negligible.
Prior to the War they had not been thought practicable
for desert use; but with their introduction it was immediately apparent not only that they were ideally
adapted to the purpose, but that, contrary to previous
beliefs, they were not confined to the use of roads and
could go cross-country in any direction with perfect
ease. The desert surfaces, especially the hammada and
the reg, offer astonishing facilities for traction, as was
early proved by the war chariots of the Pharaohs, and
in our own times by the great Boer ox carts in the Kalahari.

This does not mean that, aside from the necessities
of war, it is either safe or prudent to attempt driving
haphazard across the desert wastes. Some sort of road
is necessary; at least some definite route which is carefully charted, chooses the best ground, and is clearly
blazed with signposts, beacons and other indications of
direction, in order to allow for no uncertainty on the
part of the chauffeur, even at night. This certainly is
something of a task, but one not by any means comparable to that which would be required in the making
of a regular road. We should remark that the use of
beacons is indispensable not alone to motors, but to air-
craft as well. A plane crashing on a well-charted route
has some chance of receiving assistance, while off the
route there is little hope of escaping fatality since help
is more than likely to go astray and fail to reach the
scene in time.

Some experimentation has been necessary to select

the type of motor vehicle best suited to these routes. For a time there was much talk of tractors and caterpillar treads, and some very meritorious tests were made by Hardt and Dubreuil. But for more humble everyday needs, the tractor must be considered a rather naïve attempt. It was an idea conceived by city men, who thought of the Sahara as one great sea of dunes. And indeed the caterpillar tractor does accomplish miracles in such regions. But the present routes all carefully avoid the dunes, which are easy enough to evade; and the tractor, on the firm ground of the routes, lacks the one essential quality of the automobile, which is the greatest possible speed. Today one never sees a single tractor. Any ordinary sturdy motor car is entirely practicable, although the perfect type has six braked wheels. Cars of this type go everywhere, driven by a heterogeneous assortment of excellent chauffeurs, who form a kind of new nomadic tribe. Among them are found Frenchmen, Legionnaires, Russian exiles and other expatriates, all specializing in following these desert trails, on which they spend their whole lives.

One difficulty still remains to be surmounted, and that is the question of refueling; for unfortunately neither automobiles nor airplanes can get along without gas. There was a time not very long since when the gasoline supplies were transported across the desert by camel caravan, an obvious absurdity; but today the transport companies have established their own systems of planting stores, and filling stations with the great red columns of the Standard Oil or Shell Companies are now found dotting the routes. The main drawback at present is the exorbitant price. At Tamanrasset, for instance, gas costs 4 francs a liter. The Companies that operate services between the oases around the fringe of

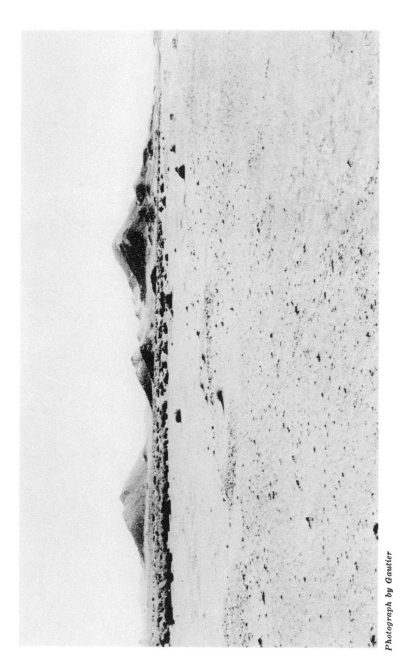

THE FIRST BROKEN STUMPS OF THE ADRAR DES IFOR'ASS SURGE UP
ABRUPTLY FROM THE REG, WHEN ONE APPROACHES FROM THE NORTH

the desert, in a proximity of 125 miles to a railway terminal, are doing very well. They have gasoline at a reasonable price and a good native clientèle. But the trans-Saharan companies which operate over distances of some 1,250 miles are not doing so well. They are failing one after the other, ruined by the high price of gas and unable to assure a regular service.

This brings up the question of trans-Saharan railroads, about which at present there is considerable debate. All over the entire world railway companies are showing a shocking deficit because they cannot compete with automobile transport; and many believe the future belongs definitely to the motor vehicle and the plane, with the steam train already a thing of the past, destined to be relegated to the museum. Partisans of the railroad reply that the question is not one of choosing between the locomotive and the gasoline motor, but between road and rail. And it cannot readily be seen how rail transportation can lose its superiority over the highway when it is a question of transporting heavy material over a long distance. Likewise, experience as well as calculations show that it should be much less costly to build a railroad on desert soil than elsewhere. In any case, the Sahara is the scene of an interesting duel between the road and the rail.

All of these innovations, together with others inaugurated by the various European administrations, have brought about astounding changes throughout the entire Sahara, felt in the different sectors in varying degrees. Egypt of course has had no need to consider the problem of trans-Saharan communications, since nature itself has there solved the problem of crossing the desert, not only by means of the Nile but by the Red Sea as well. In Egypt it is the political conditions which have undergone the most radical change. Although nationalistic Egypt obtained its independence after the War, England, who could not afford to give up her interest in the Suez Canal, took precautions to retain the ability of exerting pressure from either side, holding Egypt between two countries over which she has definite control. One is Zionistic Palestine, with which we are not here concerned; but the other is the Anglo-Egyptian Sudan, which under her rule has gone through an amazing transformation, principally economic.

Two important new industries have been instituted: the intensive cultivation of cotton with irrigation from the Nile, and the raising of gum acacias and manufacture of gum arabic, the latter of which has caused a serious repercussion upon the Senegalese trade at the opposite end of the Sahara. Naturally the modernization of communications in the Anglo-Egyptian Sudan is keeping pace with the economic evolution. The English are constructing real roads instead of mere trails, using a new process. The road bed is excavated with steam shovels and the ditch filled in with a thick layer of bitumen, a construction which is very rapid and gives a splendid surface, on which automobiles can attain a speed impossible on the trails. The only disadvantage of this method is that it is inordinately costly.

A network of railways has been completed around Khartum, the capital of the Anglo-Egyptian Sudan, connecting it with the neighboring Red Sea at the old port of Suakin, which has been brought up to date and renamed Port Sudan. Oddly enough, in spite of the fact that the cataracts of the Nile interrupt river navigation between Khartum and Egypt, the Sudanese railways have not been joined to the Egyptian system. This gap, which would be so easy to close, will probably remain open for some time, since England is not immediately anxious to complete the Cape to Cairo road and has less interest than formerly in nationalistic and independent Cairo.

West of the Nile, trans-Saharan communications have made more or less progress according to the section engaged. In Libya they are already admirably organized, in spite of the short time it has been reoccupied by Italy. The Italians have adopted without hesitation the Anglo-Sudan system of bitumen surfacing for their roads, since they have completely discarded any official use of camel caravans and depend entirely on motor transport. Italy is also pursuing a definite policy of colonization, substituting Italian for native labor and Italian chauffeurs for Arab cameleers. It is well known, moreover, that Mussolini, despite budget limitations, never hesitates when expenditures are for the imperial advantage. Consequently the bituminous roads have already been pushed into the most remote corners of the interior, reaching all such outposts as Kufara, Ghadames and Ghat.

Farther to the south, where the great Chad basin spreads out, conditions are less favorable, except for that corner which is occupied by British Nigeria where it touches the southwest bank of the lake. Nigeria of course for centuries, perhaps for thousands of years, has been a specially favored province in the black world. Around the Gulf of Guinea it has something like 70 inhabitants per square mile, with cities of from 100,000 to 200,000 inhabitants; and the city of Kano, situated almost on the very banks of the lake, has a population of 50,000. With a rich peasantry and a British administration, always commercially enterprising, Nigeria even before the War had constructed a network of railways that were pushed as far inland as Kano. It is true that the freight rates on these Nigerian railways have always, even since the depreciation of the pound sterling, been the highest in the world.

But Nigeria also has a magnificent river route penetrating into the interior. This is unique in Black Africa, where practically all of the great rivers, including the Congo, the Zambezi and even the Niger itself, are impeded near the coast by cataracts and impassable rapids. Nigeria, however, has the Benue, a great affluent of the Niger, which joins it below the rapids and opens toward Chad a water route accessible to river steamers as far as the port of Yola, some 750 miles from the ocean. From Yola there is easy communication with the basin of the Shari and the lake, so that in the present state of things Nigeria has the only open route connecting the Chad basin with the Atlantic. This open lane, it is true, ends on the south bank of the so-called lake, which we know to be only an immense morass on which navigation is impossible.

There are no other routes connecting the Chad basin with the Atlantic which have yet been developed to any extent. The former German colony of the Cameroons, now administered under French mandate, touches another corner of the lake, at which point, very near the Nigerian frontier, is found the city of Mora with 30,000 inhabitants. But this important city has never been put into contact with the ocean, as the efforts of the German authorities, prolonged by those of the French, have resulted so far in the establishment of a railroad only in the forest zone along the coast, which is a much richer region but is separated from the Chad basin by the powerful Adamawa massif with its steep slopes.

All the rest of the Chad basin, which means the central and by far the largest portion of it, is French territory. All the major trans-Saharan lanes, not only the present ones which are as yet only rudimentary trails, but those of the future, must pass through this district. This region, above the morasses of Nigeria, was formerly the field for the ravages of the Arab slave hunters, who came by way of the Sultanate of Abecher, capital of Wadaï. It was doubtless here that these raids persisted longest, since through Wadaï and the Egyptian Sudan the slave merchants found their best customers in Arabia and the Near East. It is clear that the Chad basin was practically depopulated by these raids. Nevertheless in this basin, where the sandy soil is often gorged with water, Tilho estimated that the Negroes surviving the raids maintained herds of some million of horned cattle, not to mention camels and sheep. Unquestionably this section should permit of prospective development as interesting as that of the Niger bend. The cultivation of cotton has already been attempted along the British frontier and, with the means of shipping it out by the Benue, it should have a good future.

But French Chad is only a part of the colony of French Equatorial Africa, whose capital is at Brazzaville on the Stanley Pool of the Congo River, far to the south in an entirely different type of country. This artificial affiliation is necessitated by the presence of British Nigeria, which juts in between French Equatorial Africa and the other large colony of French Occidental Africa. The two are eventually joined, it is true, but in the Sahara back of Nigeria.

French Equatorial Africa is called the Cinderella of the French Colonies because its present resources are far inferior to its future possibilities. These resources, such as they are, have naturally been held for the opening of a port on the Atlantic Ocean. Not until 1934 was the Congo-Ocean Railway opened, beginning at the new port of Pointe-Noire and ending at Brazzaville. This great economic event, which made but little stir in the outside world, was nevertheless of tremendous importance for Chad, which has natural communications with Brazzaville through a fine waterway made up of the Congo and its affluent, the Ubangi. It was for this reason that Chad was politically affiliated with Brazzaville, although the river route was only of theoretical, rather than practical, value as long as Brazzaville remained in a cul-de-sac without communication with the ocean. Now since 1934 this situation has been rectified.

Italy, under Mussolini, would like to construct a trans-Saharan railway from Tripoli to Lake Chad and thence to the ocean; and would prefer one entirely Italian. The realization of this project would require France to cede to Italy not only the Chad basin, but also the mandate over the Cameroons; but French public opinion is certainly not prepared to make any such important concession. Chad would seem to be the only portion of the Sahara which threatens to trouble European diplomacy, which has no need for even this small addition to existing complications.

Atlantic Mauretania, the most westerly part of the Sahara, is also retarded in its evolution. We must consider how isolated it is. It comes squarely up against the ancient cliffs of primary red sandstone of Hauk, Adrar, Tagant, Tishitt and Walata, beyond which in the interior stretch the great virgin solitudes of the Juf, prolonged by the abandoned Esh-Shesh Erg. The high massif of El Eglab is likewise void, and the wells of

the Wadi Shenashan have never been frequented except by some *ghazzu,* or pillaging forays, of Moroccan
Berbers. Between Atar and Chinguette on one side,
and Reggan or Wallen on the other, stretch some 930
empty miles. Mauretania has no communications with
the rest of the Sahara except by way of one trail in the
extreme north along the foot of the Atlas. According to
our latest information (1934), it is gradually becoming
motorized, and automobiles now circulate between
Tinduf and Atar; but so far these are only military
cars. It is in every respect a closed world which evolves
by itself. It has but one exceptional advantage, its Atlantic coast; and this is already under exploitation.

The fisheries of Port-Etienne have been established
in the shelter of a long sandy peninsula which extends
toward the south. In the immediate vicinity are the Ar-
guin banks, less famous than the Newfoundland banks,
but as rich in sea life; while Port-Etienne has the advantage of the dry climate and the Saharan salt to pre-
pare its fish. And it has for its market not Europe, but
all the coasts of Black Africa as far as the Congo. The
lack of cattle in this region caused by the tsetse fly has
created a meat famine so desperate that it has sometimes been cited as a justification of cannibalism; and
while the coasts of course often abound in fish, fresh
fish in such a warm and humid climate is for immediate
consumption only. Consequently Black Africa, avid
for a meat diet, finds the dried fish of Port-Etienne very
welcome. We might add that the fishermen of Port-
Etienne are mostly immigrants from Europe, many
of them from French Brittany and still more from
Spanish Galicia, all adapting themselves very well to
the dry climate, freshened as it is by the trade winds
and the ocean.

In spite of the impetus given to the experiment by these advantages, there are also great disadvantages which hinder its development. Port-Etienne has no drinkable water and must import it from Dakar in water-tankers. It is situated in the immediate vicinity of the Spanish frontier of Rio de Oro, which may be the reason why the problem of drinking water has not yet been solved, since this proximity reduces the area of exploration for an underground water supply. It certainly necessitates precautions against pillage by the Moors; for the fearsome Uled Delim are near by, and much of the time Port-Etienne is forced to live behind barbed-wire defenses.

The fisheries are not the only modern enterprise in Mauretania. There is a regular air mail service from Toulouse in the south of France to St. Louis and Dakar in Senegal, with stops at Casablanca, Villa Cisneros and Port-Etienne. This long uninterrupted stretch of coast has always been appreciated by those who fly, from migratory birds to airmen. The aviator follows the route of the storks; from the upper air he sees the coast, giving him his direction much more easily than do the beaconed routes in the interior. But there are occasional crashes, in which case the aviator is made prisoner by the Moors, and while neither killed nor tortured, is held for ransom. A very young colonel of the French Air Force once asked permission to fly the Toulouse-Dakar route; authorization was given him with the express condition that he disguise himself as a lieutenant, since, as he was told, "If you came down and were captured by the Moors, it would wreck the budget to ransom a colonel."

The Spanish base for their holdings in Africa is really in the Canary Islands; hardly any part of the

Rio de Oro is occupied except Villa Cisneros, and there are no police forces in the territory. The Moorish bandits, therefore, find a refuge in the Rio de Oro which is diplomatically inviolable; and naturally, in spite of the surveillance of the Spanish authorities, Villa Cisneros has become the center of a powerful contraband ring for smuggling arms and ammunition. The situation is also aggravated by the irregularity of the rainfall, for when the latest rains have fallen in pasture grounds controlled by the French meharists, the Moors ask pardon for their past misdeeds and promise to be good; but they instantly forget these promises when other rains have refreshed more distant pastures and they can move beyond the bounds of organized authority.

Such a situation cannot go on indefinitely. It could not possibly of course cause any real breach in the amicable relations between the two governments of France and Spain; in fact it is quite possible that some agreement will soon be reached whereby Spain will either organize a force to police the district or will accord to the French forces the right to send punitive expeditions into the country. But even if such an arrangement is not arrived at, it will require only the coördinated efforts of the three French administrations, Moroccan, Algerian and Equatorial. This of course would not be an easy matter. Nevertheless the Algerian meharists of Beni Abbes have already established outposts at Tabelbalet and Bon-Bernous; the Moroccan troops have occupied Tafilalelt, the Dra'a and even, it is rumored, the little oasis of Tinduf; while in French Equatorial Africa the old stations of Atar and Chinguette have outposts to the north such as Kediat Ijil. Thus the number of uncontrolled pasture grounds is reduced from day to day.

The most up-to-date portion of the Sahara at present, except for the Nile, is that which extends between Algeria and the elbow of the Niger: the Tuareg Sahara of the great massifs around the Ahaggar. The largest part of this is Algerian territory, which naturally has made notable advances, since Algeria for a century has been French. It was here that Lapierre created that splendid instrument of pacification, the meharist companies, and that Father de Foucauld exercised his apostolic influence. Peace and order in this great domain have been established since the first years of the twentieth century, and there has been no real trouble except briefly with the Senusi. Some thirty years of peace must necessarily constitute a major factor in the evolution of a country.

This is the region in which constant and persistent effort, enriched by experience though sometimes by catastrophe, has at length opened up the depths of the Sahara to the automobile and the plane. To get a good idea of the progress already accomplished in desert transportation, consider that in 1904-5 the author of this book spent eleven months on camel back in crossing from Algeria to the elbow of the Niger, while twenty years later he went from Burem on the Niger to Colomb-Bechar, terminal of the Algerian railway, in four days and a half. A plane makes the crossing in 24 hours. This of course is exceptional rapidity. Nevertheless there are splendid facilities for general traffic, not only for travelers but for merchandise. Routes have been established, carefully planned, well marked and kept in good condition for the requirements of motor buses, chauffeurs and tourists, with *burj,* or inns, way stations, and a few good hotels en route.

There are two main trans-Saharan routes, each filling a different need. The most easterly leaves from the railway terminal of Tuggurt, passes through Wargla, Fort Lallemand, Gassi Tuil, Flatters, the valley of the Igharghar, and Amguid, reaching Tamanrasset, capital of Ahaggar; thence by way of Adrar des Iforas, the station of Kidal, to Gao on the elbow of the Niger. This route goes through the high mountainous country, which is not only healthful and picturesque, but is the true heart of the Sahara, since it is Tuareg country.

The western route leaves Colomb-Bechar, terminal of the Oran Railway, follows the Saura and the Tuat as far as the last oasis of Reggan, and from there goes straight across the Tanezrouft till it joins the valley of the Wadi Tilemsi which leads to Gao. At Reggan there is a wayside inn conducted as a small hotel; also a signpost with a finger pointing south and the somewhat humorous though exact inscription: Gao— 745 miles. The next lap is indeed 745 miles, without a watering place or a stop. It is a somewhat rash undertaking, but has been justified with success. For the automobile, unlike the camel who must eat and drink, has no need to worry about pasturages or consider anything except the straight line which is the shortest distance from one point to the other. [See illustration, p. 234.]

This crossing is marked at various points by large gasoline tins, empty of gas and filled with sand. Each has a number, and there is one about the middle of this terrible route which bears the number 5. This number 5 marks a spot which has acquired a certain specific notoriety. The transport company at this place went to some trouble to install a little filling station where gas and water might be procured, and left a Negro in charge of it. The man, a pathetic victim of speculation, was found dead of thirst, because he had sold his own pro-

vision of water to passing motorists, tempted by the
high price they offered.

Everything on this westerly route is sacrificed to
speed, for it is used only by people who are in a hurry;
who have no business in the Sahara itself, but wish to
reach the elbow of the Niger as quickly as possible. Between the two main routes there are a number of intermediate routes serving the oases: In Salah in Tidekelt, El Golia, Timmimun in Gurara, and Laghwat
which has a fine road connecting it with Jelfa and
Tozeur, terminals of the Algerian and Tunisian railways.

There are also branch lines from these two which
serve the Chad region. One leaves from Gao, follows the
Niger down as far as Nyamey near the British frontier,
and thence along this frontier by way of Zinder to
Nguigmi on the lake. This makes a considerable detour.
Another, which is much more direct, has just been
opened, leaving from Tamanrasset and going by way of
the wells of In Guezzam to Agades and thence to Zinder.
This route is particularly useful for the international
air mail lines; the French line will presently follow it
in joining those of the Belgian Congo which are already
functioning.

This Tuareg Sahara is particularly the center of the
controversy regarding the feasibility of trans-Saharan
rail transportation. Construction on one rail project
is about to begin, the plans being already carefully
studied. The road will leave from Colomb-Bechar, terminal of the Algerian and Moroccan railways, and will
follow closely the route of the occidental road as far as
Gao or Burem; there it will divide, one branch going
upriver along the Niger as far as Kulikoro, terminal
of the Dakar railway. The other will go down the

Photograph by Gautier

VENTILATION VENTS OF THE FOGGARAS AT REGGAN, LOWER TUAT

Niger as far as the British frontier and join the Nigerian system.

The project should be easy of realization, for the route is charted in French territory, which is perfectly peaceful; and the execution should take only a few years. The total cost is estimated at 3 billions of francs, which is a comparatively small sum in the annual French budget of 50 billions. At present only the vote of Parliament is lacking, but public opinion is indifferent on the subject, or perhaps even a little hostile. This is because there is not sufficient confidence in the economic future of the Niger region; and the reason is that modern France, unlike America—or at least the America of yesterday, lacks the hopeful spirit. But in spite of the wordy arguments which hamper present progress, life will go right on, and deeds will presently be accomplished.

Certain economic consequences of the new era have already made themselves noticeably felt. One phase which is particularly striking is the decline of the old Saharan industries. The salt traffic of the Sahara, that of Taudeni for instance, can no longer monopolize the market in the Sudan, for competition has been introduced by the importation of European salt from across the sea. European cloth stuffs are causing the native trades to disappear one by one; and since the ostrich has been raised in South Africa, the caravan traffic in Sudanese plumes has lost all importance. Likewise the gum-arabic trade of St. Louis in Senegal has suffered a collapse from the sudden competition offered by the British product from the Sudan, where England of course with her energy and experience can easily outdistance her competitors and capture the market. The Saharan gum, raised for centuries by the Moors at

Trarza, north of the Niger, and shipped to St. Louis, has been manufactured by primitive and traditional methods, all the more difficult to modify because they are so ancient. All these factors form one small aspect of the present world economic crisis.

But what most of all has given the truly mortal blow to the old trans-Saharan commerce has been the disappearance of slavery and the suppression of the slave trade. For the real basis of this commerce throughout thousands of years was the black slave destined for Egypt and the Maghrib. Consequently the Saharan oases which were formerly great commercial centers, such as Aïr and Fezzan, are now in complete decadence. The nomads too have felt the effects of this depression, for as sovereign overlords of the great trade routes they used to levy heavy tolls, which they now find very much diminished. Their natural tendency to pillage is increasing; and, as always happens, indigence and insecurity continue to enhance one another, forming a vicious circle.

There is a certain measure of compensation wherever a European military occupation is introduced. The garrison draws regular pay and has nowhere to spend it except on the spot. The oases of the French Sahara live on their garrisons, and this one small economic factor contributes enormously to the political tranquillity of the country. European industry is also beginning to creep in along various lines, sowing the seeds of a new prosperity. Egypt is overflowing with gold since the French have opened the Suez Canal and the English have organized the cultivation of cotton in the valley of the Nile. In the French oases of the Lower Igharghar a new impetus has been given to the cultivation of the date since machine methods have multiplied the number of artesian wells, and this has been augmented by the generally higher prices of alimentary products since the War. At Kenatsa, on the edge of the Sahara along the frontier between Morocco and Algeria, a tiny seam of coal has been under exploitation for some years. And in such a country as this, where the population is so sparse and so close to destitution, it does not take a very great financial enterprise to restore the balance between production and consumption.

Once begun, the movement will continue. It is remarkable that as yet no new artesian region has been discovered, for there are many places under the immense rolling plateaus of hard calcareous or sandstone rock where the ideal conditions for artesian fields would seem theoretically fulfilled, and a scientific study should reveal them with precision. Nor would the hardness of the rock constitute an obstacle to the up-to-date machine, as it did to the native pickax. As for mineral resources, a region as ancient as the Sahara does not of course offer the same possibilities as newer fields like Australia, California and Alaska. Any surface gold, for instance, that there may have been, was long since scratched out, gathered up and drained off to the profit of the early Mediterranean civilizations. Nevertheless we can hardly believe that in all these vast spaces, which take in half a continent, there is not some interesting prospect of mineral development remaining for the future. As for agricultural possibilities, we have the region bordering on the Sudan, especially at the bend of the Niger, where it will manifestly take little more than the will to transform the great empty expanses into flourishing garden lands.

We must naturally guard against exaggeration. The finest desert in the world will never in its entirety lend

itself to any wholesale exploitation unless man dis-
covers the secret of making rain; though we may certainly expect some miracles of science only a little less
unlikely than this. It will undoubtedly one day be possible to draw some power from the two great Saharan
forces of solar energy and wind which now go to waste;
but, even with this, the Sahara will remain always the
Sahara.

Nevertheless there are mighty forces at work which
seem about to lead the desert into new ways. The total
figure of the present population cannot be determined,
but we know for a certainty that it is insignificant, that
this half of the continent is practically empty. This in
one way is a difficulty, but in another is a great advantage. For Europe does not find here a dense and
deeply rooted native population with which it must
cope. And the European also finds here a climate which
has been demonstrated by practical experience to be
perfectly adapted to the Mediterranean white race.

As for the occidental portion of the desert, it seems
certain that the success of the French policies and colonization in the Maghrib will inevitably be carried on
into the Sahara. Something must be done to establish
a bond between the colonists and the natives which will
include both sentiment and enterprise; they must be
associated in the undertaking of some great common
task. Such a task is offered in the Sahara as nowhere
else in the world.

The desert's best chances for the future are determined by its location on the globe. It is interposed
between two vast regions of violent contrast which have
a great need and a strong attraction for each other. To
one side of it we find the civilized agglomerations of
Western Europe, and to the other the unexploited agri-

cultural wealth of the African tropics. Europe since the War has felt more than ever the need to exchange its manufactured products for food products. Then too, we must allow for the imaginary need for new landscapes of our great tourist epoch. Europe, if we may put it so, is famished for her tropics, from which she has always been separated by the obstacle of the Sahara. But in our day, with the annihilation of space, such an obstacle is absurd and must be done away with. There is a real necessity. We feel already that the impetus has been given, and that the Sahara is starting a new chapter in its history.

Glossary

Ammon Ra. The ram-headed God of Thebes. His cult was disseminated through the desert by way of the "oasis route" and Siwah.

Ashab. An ephemeral desert vegetation developing after the rains. It is the favorite pasturage of the camels.

Azalai. The great caravan which comes periodically from the salt works of Taudeni to Timbuktu.

Bahr. (Arabic term signifying *the sea;* plural, bahar.) A deep natural spring, often in the form of a small crater-shaped lake of great depth. Such springs are found in some of the eastern Algerian oases.

Balad-al-khuf. The country of fear.

Balad-as-sif. The country of the sword.

Barkhan. A small crescentic wind-formed dune.

Burj. An inn.

Daglat nur. A choice variety of date grown in the eastern Algerian oases at the foot of the Atlas.

Daya. A depression in a limestone or sandstone plateau, lined with vegetation dependent on subterranean circulation.

Dhu-l-Karnain. (*Two-Horned.*) Alexander the Great as the incarnation of Ammon, credited with being the founder of the artesian wells of Wadi R'ir.

Dillu. A leather bucket with a long handle, used for irrigation in the western Algerian oases.

Erg. (*Sandy desert.*) A vast region covered deeply with pure sand, occupied by dunes. Erg-el-Atchan and Erg-er-Rawi signify, respectively, *erg of thirst* and *humid erg.*

Fejj. (*Neck, closed ground.*) A long sand-free passage in an erg region.

Foggara. An oriental irrigation system consisting in long underground aqueducts for storing water. Foggaras are found in Baharia and Farafra in the Libyan Desert, and in the western group of Algerian oases.

Gassi.   A long sand-free passage in an erg region; a fejj.

Gharb.   The western half of the Moslem world.

Ghazzu.   A pillaging foray.

Gulglan. (Arabic.)   A particular variety of *ashab* vegetation: *Savignya longistyla,* a rosette plant bearing purple flowers.

Hammada. (*Rocky desert.*)   A tableland or plateau of rock denuded by wind erosion.

Haratin.   The Negro cultivators of the Algerian oases.

Harmattan. (Native name.)   The desert wind.

Khammis. (Arabic, *five.*)   A term applied to Negro cultivators or haratin in Algerian oases because of their right to a fifth part of the produce of the land.

Khamsin. (*Wind of fifty consecutive days.*)   In the Egyptian desert, the wind from the southwest.

Kharafish.   In the Libyan Desert, a type of limestone plateau formed by wind erosion, analogous to the *yardang* of inner Asia.

Khottara.   A primitive device for raising the water of natural wells, found in the western Algerian oases.

Ksar.   A fortified town of the Algerian Sahara.

Ksurians.   The inhabitants of the *ksars.*

Kuffin. (*Bale.*)   A part of the primitive equipment of native well makers in Algerian oases.

Litham.   The veil worn by the Tuaregs and Tibbus.

Mehari.   A riding camel.

Meharists.   Mounted camel troops.

Mushrabias.   Lattices of fretted wood.

Oghurd.   A massive mountainous dune formed by some underlying rocky typographical feature.

Rahla.   A camel saddle used in the western desert.

Reg. (*Stony desert.*)   A type of desert surface consisting in beds of pure gravel of alluvial origin from which the the sand has been removed by wind. It is the formation prevalent in the western Sahara.

Sabkha.   The terminal basin of a desert wadi, wherein salt is deposited by evaporation.

Serir. (*Stony desert.*) A formation analogous to the *reg* but much older. It is made up of pebbles and rounded stones instead of gravel.

Shabka. (Arabic term signifying *network* or *fiber*.) A type of desert landscape formed by wind erosion of alluvial basins.

Shaduf. A device similar to a *khottara,* for raising the water of natural wells. It is used in Egypt.

Shahad. Moslem funeral stela.

Shahali. (Native term signifying *wind from the south.*) The desert wind in the Sahara proper.

Shark. The eastern half of the Moslem world.

Shott. The terminal basin of a desert wadi, wherein salt is deposited. The formation is similar to the *sabkha.*

Sif. (*Sword.*) A long, slightly incurving, ridged dune.

Simoom. The desert.

Sirocco. The desert wind.

Tanezrouft. A region of maximum aridity.

Tifinar. The ancient Libyan alphabet, now used only by the Tuaregs.

Uran. A great lizard.

Wadi. A desert watercourse ending in an alluvial region in a closed basin.

Zagaya. The casting weapon of the Berbers.

Index

A

Abecher, capital of Wadaï, 227

Ablutions, avoided by Tuaregs for fear of stimulation of sweat glands, 16

Aborigines, 132, 198, 212

Abu-Moharique, perhaps formed by wind action, 49-50

Abydos, 155

Abyssinia, Egyptian and Roman influence upon, 143-44

Abyssinian range, 65; Nile fed by springs of, 73

Acacias, 154, 224

Adamawa (massif), 65; Cameroons separated from Chad by, 226; Shari fed by streams from, 68

Adax, desert inhabited by, 23

Adrar, Mauretanian, see Mauretanian Adrar

Adrar des Iforas (massif), 233; altitude of, 181; Kuntas reach slopes of, 209; population of, 206-7

Africa, see Black Africa; Central Africa; East Africa

Agades, 207, 234

Agriculture: dependence upon communications, 155; experimentation in region of the Terres Sialines, 96; future possibilities of, 237; grazing incompatible with, 154

Ahaggar (massif), 31, 114; counterpart, in Occidental Sahara, of Tibesti, 161; description of, 181; fossil wadis in, derivation of, 56; Igharghar, waters of, furnished by, 90; map of, 182; Mount Ilaman, summit of, 31; Quaternary rains, traces of, found in, 63; Tibesti, relation to, 173; volcanic, 31; wadis originating in, 88

Ahaggar Tuaregs: conflict of Ajjers with, 206; Iforas dominated by, 207

Ahenet (massif), 211; Ahaggar Tuaregs graze camels in, 206; altitude, 181

Aïn Dua, ochre paintings in, 131

Aïn Galaka, dates grown at, 166

Aïr (massif), 31, 187, 206; altitude of, 181; effects of erosion upon, 40; Hausas in, 127; human life in, 207-8; importance of slave trade to, 236; source of wadis in, 88; Tuareg conquest of, 132; volcanic, 31

Air mail: route to Chad, 234; Toulouse-Dakar route, 230

Airplane: Algerian Sahara open to, 232; use of, tested in World War, 221

Ajjer Tuaregs, 97; conflict of Ahaggars with, 206; Ghat dominated by, 187

Alexander the Great, 158, 189

Algeria, 32, 184; Cretaceous limestone overlay in, 30; French administration of, 232; nomads of, 149; plant life exists in, 19; sandstones of, Cretaceous, 55

Algerian Geological Service, 54

Algerian Military Territories, 32

Algerian plateaus: big game country, 22; fossils found in, 54; steppes, 18

Algerian Sahara: conquest of, by white race, 129-31; fossils found in, 54; geology of, 29-30; Igharghar lies entirely within, 89; inhabitants, 184-99; map of, 92; pacification of, 202-3; regs of, work of fossil wadis, 62; wadis and ergs in, 102-3; wells and pasturage in ergs of, 98-99. See also Occidental Sahara

Alluvial deposits: ergs caused by eolian erosion of, 101-6; wadis' base level determined by, 39; winds counteract accumulation of, 37. See also Reg; Sedimentation; Serir

Alluvial regions: Igharghar, 91-93; wadis terminate in, 36; White Nile spreads in, 72-73

Almoravides, Sanhajas identical with, 213

Alphabet, Libyan, used by Tuaregs, 204

Alps, Atlas range comparable to, 29

Civilization: Egypt, ancient center of, 139; sedentary peoples contribute most to, 151

Clarias lazera, tropical fish common in terminal region of Igharghar, 58-59

Cleanliness, rites of, superfluous in desert, 16

Cliffs: absence of, evidence of absence of lake with fixed level, 58; geology of, at Siwah, 157

Climate: desert, changes in, 94-97; desert, stability of, 53; desiccation of Sahara not due to changes in, 105; function of latitude, 6-7; Saharan, adapted to white race, 238; Saharan, description of, 9-17; shape and height of land, factors in, 7-8; substitution of camel for elephant may be due to changes in, 125. See also Aridity; Rainfall

Closed basins: antiquity of, 54; desert topography characterized by, 36-37; salt marks presence of, 99; wadis have base level in, 38. See also Bilma; Igharghar; Taudeni

Cloth, European, importation of, 235

Coal, seam of, at Kenatsa, 237

Coasts, see Seacoasts

Cobras, Saharan, degeneration of, 123

Cold, see Temperature

Colomb-Bechar, 232, 233; railroad terminus, 234

Colonization: conditions created by, 134; economic effects of, 235-37; Saharan, 219-39

Colossus of Memnon, 118

Comité de l'Afrique Française, 161

Commerce: between Leptis Magna and Black Africa, camel introduced to facilitate, 133-34; early trans-Saharan, Ghadames connected with, 186; Red Sea and Mediterranean as carriers of, 140-41; Sahara a barrier to, 25

Communication: improvements in, 220; problem of, inseparable from that of agriculture, 155

Congo (river): cataracts, 226; waterway to Chad, 228

Congo-Ocean Railway, 228

Constantine Tell, fossils found in, 54

Coral, 141

Corippus, 124, 189, 204, 207

Corrasion: appearance of shabka partly due to, 104; obstacles checking effect of, 42-44. See also Eolian erosion

Cortier, M., 215

Cotton: cultivation of, in Chad, 227; cultivation of, in Nile valley, 236; Sudanese, 224

Cretaceous limestones, see Limestones

Cretaceous sandstones, see Sandstones

Crocodile, 143; degeneration of, 123; found in Tibesti water holes, 88, 162; immigrant from the tropics, 59-60

Croûte calcaire, cause and effect of, 51-52

Cycles, desert erosion, 104

Cydamus, see Ghadames

Cyrenaica, 32, 112, 115; Cretaceous limestone overlay in, 30; Jewries of, 129; southern boundary of, 160; steppes in, 18

D

Daglat nur, see Dates

Dakar: Geological Service of, 211; water imported for Port Etienne from, 230

Dakhla, 62; geology of, 152; population of, 155; Senusi capture of, 156; water at, 153

Dalloui expedition, Tibesti explored by, 161

Darius, 156

Dates: Algerian, 191-92; Borku, 166; cereals replaced by, 128; cold spells necessary to growth of certain varieties of, 13; Fezzan, 177; Igharghar alluvial region, source of, 91; impetus to raising of, new, 236-37; Sudanese atmosphere too wet for, 206

Dayas, description of, 99-101

Dead Sea, 74

Death, danger of, in the Sahara, 116-18, 201

Depressions, system of, runs from Cairo to In Salah, 32

Deserts: age of, tanezroufts due to, 113-14; climate in, stability of, 53-54; closed basins, a structural trait

J

Jackals, vicinity of Wallen well inhabited by, 24

Janet, fault at, 187

Java, wood imported from, 142

Jebel es Soda, see Black Mountain

Jebel Nefussa, 185

Jedi (wadi), course of, 91; former frontier between Berbers and Ethiopians, 126

Jelfa, 234

Jenne, 66; Timbuktu, outpost for, 210

Jerabub, 32; joint-capital of Senusism, 160

Jerat (wadi), paintings and rock carvings found in vicinity of, 97; rock carvings in, 131

Jerid: artesian wells at, 190; oases of, 185

Jerid (wadi), dates grown in oases of, 91

Jerma-Garama, 124, 178, 186; antiquity of, 174

Jews, see Zenetes

Jinns, 16, 40, 118

Juf, 228; former terminus of the Upper Niger, 66-67; tanezrouft, 113

Jupiter Ammon, 156, 157

K

Kalahari, 99; antiquity of, 55; classification of, as steppe, 18; desiccation, evidence of, found in, 97; ox carts used in, 221; ruminants, once overrun with great herds of, 22

Kano: population of, 225; railroad to, 226

Kanuri, 178; Tibbus related to, 164

Kawar, 185; mining of salt at, 179; Tibbu territory, 170

Kediat Ijil, outpost at, 231

Kel-Geress, description of, 208

Kenatsa, coal mined at, 237

Ketamas, Fatimid Empire established by, 214

Khammis, see Haratin

Khamsin: characteristics of, 16; opaqueness of, 45; origin of, in erg of Libyan Desert, 15

Kharafish, 113

Kharga, 62; geology of, 152; military importance of, 156; population of, 155; water at, 153

Khartum, 32; railway reaches to, 144, 224

Khottara, use of, in irrigation, 194-95

Kidal, 233

Kiliau, Conrad, 187

Knives, see Throwing knife

Koran, 189

Kosseir, 142, 155

Ksars: description of, 197; fortification of, 199

Ksurians: Negroid, 198; servility of, 199-200. See also Haratin

Kudia, see Atakor

Kufara, 112, 115, 188; conquest of, by Arabs, 132-33; description of, 168-70; joint-capital of Senusism, 160; roads to, 225

Kulikoro, 234

Kuntas, description of, 209

Kusaila, 207

L

Laghwat, 91, 234

Lakes, Edeyen and Igharghar Ergs contain, 176. See also Bahr-el-Dud; Chad; Wau en Namus

Landscapes, Saharan: appear strange, 34; result of eolian erosion, 41-52

Laperrine, Col. Marie-Joseph-François-Henry, 4, 116, 117; death of, 117

Lapierre, Sergeant Major Laurent, 5, 168, 169, 170; meharist companies established by, 232

Laterite, found in Tibesti, 162

Latitude, climate a function of, 6-7

Latrines, 196

Lava fields, 109, 181

Lavauden, L., 107; evidence of Ahaggar desiccation found by, 96

Learning, attribute of sedentary peoples, 151

Legends: danger of emotion personified in, 117-18; Iforas, 207

Lenz, O., 215; exploration of Mauretania by, 211

Leptis Magna, 124, 126, 133

Leuke Kome, see Kosseir

Printed and bound by CPI Group (UK) Ltd, Croydon, CR0 4YY

23/10/2024

01778232-0004

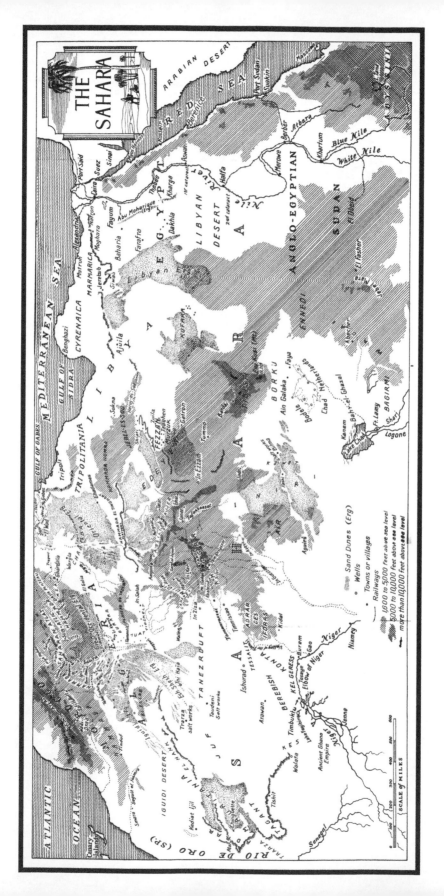